Contents

About the Authors

Victor Dmitriev is Doctor in Electrical Engineering, a full professor at the Department of Electrical and Biomedical Engineering, Federal University of Pará, Belém, Pará, Brazil. Victor Dmitriev received B.Eng. degree in 1971 in the area of high energy systems, and Ph.D. in 1977 in the field of microwave devices, both in Bauman Technical University, Moscow, Russia. From 1977 until 1994, he teached at this University. In this period, Prof. Dmitriev participated actively in many scientific projects related to radar technology. In 1980, he worked as a visiting professor at Manchester University (UMIST, Great Britain) investigating theoretically and experimentally microwave ferrite devices. Since 1997, he has been a professor at the Federal University of Pará (UFPA), Belém, Brazil. His research interests include group-theoretical methods in electromagnetic wave theory, nanophotonics, nanoelectronics, metasurfaces, nanoantennas, and partial discharges. He published over 150 journal papers, 350 conference papers, 45 patents, 10 books, and 9 book chapters.

Rodrigo M. S. de Oliveira has a Doctoral degree in Electrical Engineering obtained in 2008 from the Federal University of Pará (UFPA). Currently, he is an associate professor at the Institute of Technology, UFPA (Belém, Pará, Brazil). He has authored and co-authored about 45 journal papers and 100 conference articles regarding development of novel numerical methods in electromagnetics, application of computational electrodynamics to model complex structures involving partial discharges, grounding systems, electromagnetic compatibility, antennas, nanotechnology, and on applying and developing optimization techniques. He is the head of Laboratory of Electromagnetics (LEMAG-UFPA), where he develops his research and advises Doctoral and Master students.

Ronaldo F. Zampolo received the bachelor's degree in Electrical Engineering from the Federal University of Pará (UFPA), Brazil, in 1995, and the M.Sc. and D.Sc. degrees in Electrical Engineering from the Federal University of Santa Catarina, Brazil, in 1998 and 2003, respectively. He joined the Department of Electrical Engineering of the UFPA in 2004. He is currently a professor at the Department of Computer and Telecommunications Engineering, UFPA and a member of the Signal Processing Laboratory (LaPS/UFPA). His current interests include partial discharge analysis, automated visual inspection, deep learning techniques, and human activities recognition.

Paulo R. Moutinho de Vilhena is Doctor in Electrical Engineering and Engineer in Eletrobras Eletronorte, Belém, Pará, Brazil. He received a technical degree in Computer Technology from the Federal Institute of Education, Science and Technology of Pará (2002), B.Eng. in Electrical Engineering from the Federal University of Pará (2005), and a specialization in Systems Engineering from the University Center of Pará State (2005). He obtained an M.Sc. in Electrical Engineering (2008) and a Ph.D. in Electrical Engineering (2015) from the Federal University of Pará. He is currently an Electrical Maintenance Engineer for Power Plants in Northern Brazil—Eletrobras Eletronorte. He has experience in the field of electrical engineering with emphasis on electric power systems.

Fernando de Souza Brasil is Doctor in Electrical Engineering and an Engineer in Eletrobras Eletronorte, Belém, Pará, Brazil. He received B.Sc. in Mathematics from the State University of Pará (2005), B.Eng. in Electrical Engineering from the Federal University of Pará (2006), specialist title in Industrial Engineering from the State University of Pará (2012), and his Master and Doctor degrees in Electrical Engineering from the Federal University of Pará in 2013 and 2016, respectively. He is currently an Electrical Maintenance Engineer at the Electric Power Plants of Northern Brazil-Eletrobras Eletronorte and Professor of Electrical Engineering at FACI Wyden. He has experience in electrical engineering with emphasis on electrical power systems, working with predictive maintenance of high voltage equipment in substations and hydroelectric plants. He is an Acoustic Emission Analyst—Level 1, and the National Coordinator of the Rotating Machine Studies Committee CE A1 (Cigre).

Martim Felipe Fernandes received his Electrical Engineering degree from the University of São Paulo (São Paulo, Brazil) in 2011. After working at Siemens in São Paulo for 3 years, he continued his studies at the University of Surrey (Guildford, United Kingdom), receiving a Masters degree in Nanotechnology and Nanoelectronic Devices in 2014. Since then, he has worked as Electrical Engineer at Eurovolt Energy Service, a company focused on predictive maintenance of large electrical machines. In 2020, he was accepted to the Ph.D. program in Electrical Engineering at the State University of Londrina (UEL) to investigate the relationship between partial discharges and ozone in rotating machines. His main research interests are asset monitoring, partial discharge analysis, data mining, predictive maintenance, and nanotechnology.

Acronyms

AC	Alternating current
ADC	Analog-to-digital converter
AM	Amplitude modulated
ANN	Artificial neural network
ANSI	American National Standards Institute
AWGN	Additive white Gaussian noise
B.I.L.	Basic impulse level
CV	Cross-validation
DOP	Damped oscillatory pulse
EB-PVD	Electron-beam physical vapor deposition
FDTD	Finite-difference time-domain
FIR	Finite impulse response
FRT	Fast rise time
GGB	Generator guide bearing
GUI	Graphical user interface
HF	High frequency
HMOS	Heated metal oxide sensors
HPP	Hydroelectric power plant
ICA	Independent component analysis
IEC	International Electrotechnical Commission
IEEE	Institute of Electrical and Electronics Engineers
IIR	Infinite impulse response
LED	Light-emitting diode
LF	Low frequency
LPA	Log-periodic antenna
LTI	Linear time invariant
MLP	Multilayer feedforward
NHL	No hidden layers
NPS	Number of pulses per second
NQN	Normalized quantity number
OSHA	Occupational Safety and Health Administration

PAS	Photoacoustic spectroscopy
PB	Passband
PCA	Principal component analysis
PD	Partial discharge
PHA	Pulse height analysis
PMA	Pulse magnitude analyzer
POT	Pulse occurrence time
PRPD	Phase-resolved partial discharge
PRR	Pulse repetition rate
RF	Radio-frequency
RFID	Radio-frequency identification
RTD	Resistance temperature detector
S/C	Semiconductor coating
SAW	Surface acoustic waves
SB	Stopband
SGD	Scaled conjugate gradient backpropagation algorithm
SNR	Signal-to-noise ratio
SSC	Stator slot coupler
SVM	Support vector machine
TGB	Turbine guide bearing
TO	Transistor outline
UHF	Ultra high frequency
UV	Ultraviolet
VET	Voltage endurance test
VHF	Very high frequency
VNA	Vector network analyzer

Chapter 1
Introduction

1.1 Rotary Machine Failures

There are two ways to classify faults in high voltage generators. One is the condition in which the machine was found after the failure and the other is due to the suspected root cause that caused the failure [1]. For diagnostic purposes, the most important is whether the development of failures would be on-line predictable from some measurement techniques.

The distribution of faults and causes in the components varies between different types of machines, but a general indication is given below. A detail study in causes of failure in hydrogenerators after examination of 69 cases of incidents [2] shows that the main causes of failures can be categorized as follows:

- Failures in the insulation system;
- Mechanical defects;
- Badly done semiconductors coating;
- Thermal problems;
- Failures due to bearings.

Figure 1.1 illustrates that the principal cause of failures is the problems in electrical insulation system.

1.2 Partial Discharges in Hydrogenerators

Unscheduled machines stoppages caused by failures in power hydrogenerators are a huge problem because to turn off the equipment to do maintenance is very costly. The predictive on-line maintenance tests have become an effective tool in the management of the production activities of generators. In this model, the

V. Dmitriev et al., *Partial Discharges in Hydroelectric Generators*, Power Systems,
https://doi.org/10.1007/978-3-031-36604-8_1

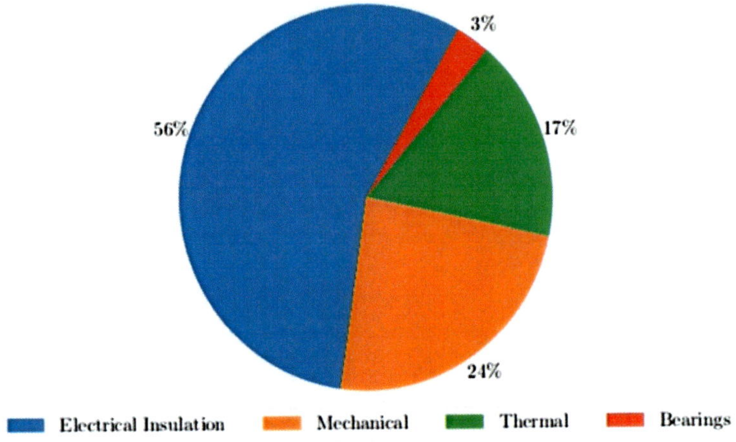

Fig. 1.1 Root causes of hydrogenerator failures

disconnections can be performed in a scheduled manner basing on the equipment conditions.

With the purpose of solving this problem, researchers studied those failures which were related to the increase of PD level of the machine electrical insulation. The PD occurrences in the insulating systems of the high voltage devices are fragility symptoms in the dielectrics, whose evolution may bring severe consequences to the equipment. Over the years, the monitoring of PD has become the more largely used method to determine the electrical insulation conditions in the stator windings of generators. Compared with other dielectric tests such as measurements of dissipation factor or insulation resistance, the PD measures allow to detect weak points in the insulation and identify them [3]. The benefits of on-line testing allow for equipment analysis and diagnostics during normal production. Corrective actions can be planned and implemented, resulting in reduced unscheduled downtime.

In the past 20 years, the main focus of the research was expanding the PD technology to be applied to diagnostic tests to determine the insulation condition of the equipment in operation. State evaluation may include on-line tests of PD combined with the off-line testing. The on-line tests demand the development of better ways for suppressing noises which can lead to false indications, and development of tools for PD interpretation.

The advances in the PD measurement technology allowed to reach the point where the off-line and on-line PD measures in-field are now routinely applied to generators. The impact of these advances is the improvement of reliability of high voltage devices since equipment degradation can be identified and the equipment repaired or substituted before catastrophic failure in service. It is important to note, however, that such diagnostic tests can not give a completely trustable indication of remaining life. The objective is to alert about an imminent problem and perhaps to identify the cause and to isolate the problem [4].

The PD measures can detect weak points in the insulation system and degradation process which include [3]:

- Poor impregnation with epoxy;
- Badly done semiconductors coating;
- An insufficient spacing between coils in the area of coils interlacing;
- Loose coils in the slot;
- Overheating (long time thermal deterioration);
- Winding contamination by the humidity, oil, dirt, etc.;
- Problems of load cycles.

Insulation quality in high voltage machines is the key question. The high levels of the electric field due to the operation of these machines promote the continuous degradation of dielectrics. This degradation provokes emergence of heterogeneity in the insulating material, in a way that the high voltage in those regions gives the origin of PD. Besides, discharges may emerge due to strange corps in the system (as screws, splinters, etc.) what justifies the effort in the knowing of the state condition of the dielectric material of this equipment.

Generically, a system to evaluate the state of dielectrics of the hydrogenerator stator windings using the electrical method of PD measurement is composed of conventional sensors and/or electromagnetic sensors, measurement circuit, acquisition and processing system providing denoising and statistical maps. Figure 1.2 illustrates the system.

The most frequent failures are provoked by the effect of aging and contamination of winding by dust and humidity. Electrical failures are caused also by internal partial discharges and overvoltages. Due to the mechanical vibration the bars suffer loosening in their position inside the slot in the coil head. Possible faults in the electrical insulation system are presented in Fig. 1.3.

1.3 Electromagnetic Sensors for Partial Discharge Measurements

PD are associated with electromagnetic emission, acoustic emission and some chemical reactions such as formation of ozone. To measure PD in practice, the electrical methods are commonly used. The simplest and most frequently utilized are the sensors called capacitive couplers [5, 6]. They are installed along the stator with direct connection to the windings (see Fig. 1.2). The sensibility of these sensors is limited mostly by their relatively narrow frequency band. The sensibility of the capacitive couplers can be enhanced increasing their capacitance to values higher than typical 80 pF. In work [7], theoretical and experimental analyses were related to the bandwidth of the sensors. The couplers that use the 500 pF capacitors can register more discharges in comparison to those based in 80 pF couplers.

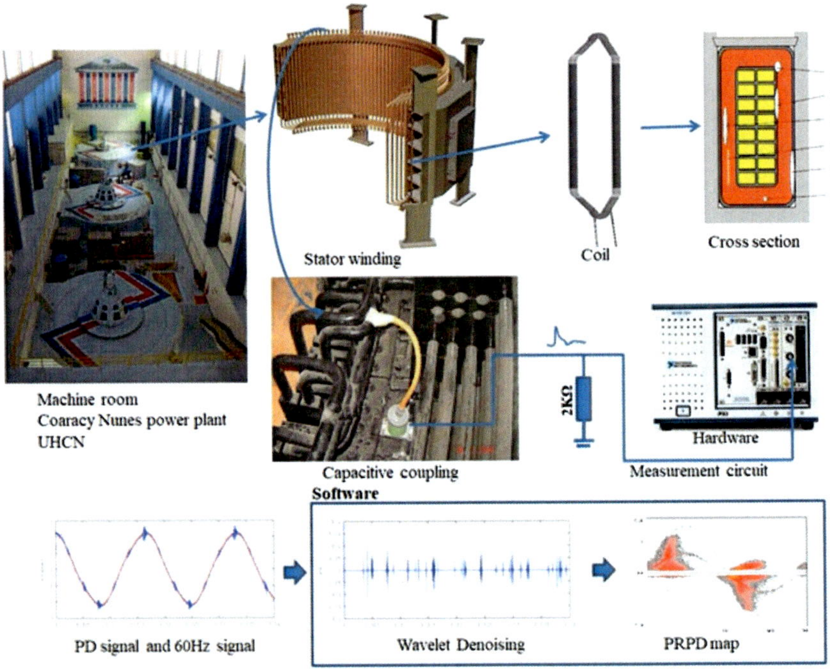

Fig. 1.2 A panorama of PD problem. From the left to the right, the first row is photograph of a hydrogenerator: a stator winding, a stator coil and its cross-section (the letters in the cross section denote typical defects and discharges: A is corona in the coil heads, B and C are delaminations, D is a region without mica layers, E is a slot discharge, F is a cavity). The second row shows a capacitive coupler and the measurement device (hardware). The third row presents typical graphics produced by the monitoring system: 60 Hz with imposed PD signals, the PD signal after noise suppression and PRPD (phase resolved PD) map after processing by software

Another type of sensors, namely, stator slot coupler (SSC) can also be used for detecting PD [8]. Unlike traditional couplers that present high voltage capacitors, the SSC is not connected directly to the winding. The SSC is a device which detects the electromagnetic energy of PD and other signals in the range of 10–1000 MHz. The SSC is installed in the stator winding slot containing stator bars [9]. The SSC presents a microstrip line, and consists of a ground plane and a tape sensor line with coaxial cable in the output. When an electromagnetic wave generated by a PD pulse spreads along the stator bar, the SSC which is close to the bar, will capture this pulse and send it to the input of a measurement device [8].

The characteristic impedance of SSC, usually of 50 Ω, is matched with the impedance of output coaxial cable. These cables can be connected to the two sides of the coupler. This allows determining the propagation direction of the PD pulse and simplifies its location. The SSC can be installed in a generator only when the rotor is removed from the machine. In [10] is proposed the use of directional couplers to register partial discharges in operating rotating machines. Different types of antennas also can be used as PD sensors as will be discussed below.

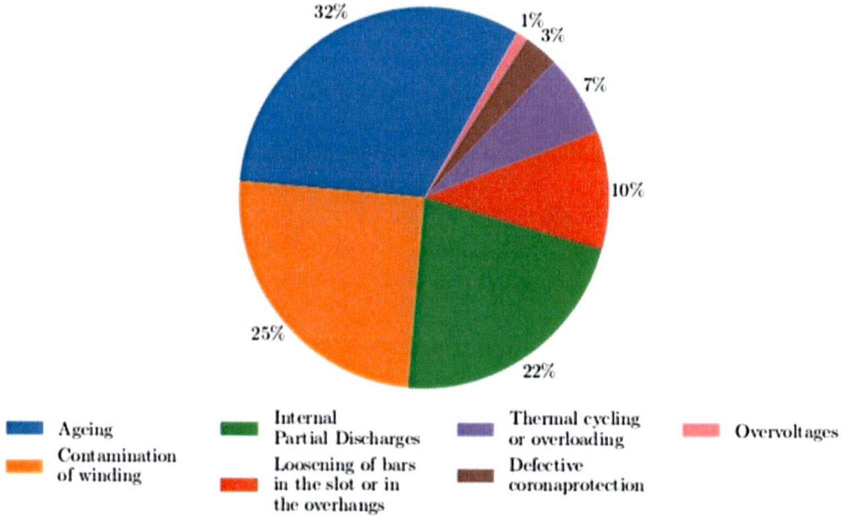

Fig. 1.3 Failure causes at electrical insulation system

1.4 Numerical Modelling and Pinpointing of Partial Discharges

Determining the precise location of partial discharges in a hydrogenerator is a challenging topic. It is not a straightforward task in a single or in a couple of stator bars either. The challenge is associated to the fact that partial discharges excite high frequency electromagnetic waves, which propagate in the stator bar insulation, reflects on its conductive parts and on the bar ends. Furthermore, electromagnetic waves are radiated from the bar to air and reflects on neighbor structures, producing multiple electromagnetic couplings. As a result, complex transient electromagnetic signals (voltages of currents) are registered using sensors.

Despite the complexity of the obtained transient signals, it is possible to identify dominant resonance frequencies associated with bar dimensions and bar composition affecting wave velocity. Below, we discuss a methodology based on resonance analysis which allows one to locate PDs on a single or in a couple of connected stator coils. Concurrent PDs are also possible to be pinpointed. The problem is analyzed numerically by using the three-dimensional finite-difference time-domain method (FDTD) [11, 12], employed to solve Maxwell's equations. FDTD calculations are used to develop the PD pinpointing method. Numerical results are also validated by comparing FDTD transient results with signals obtained in laboratory experiments.

1.5 Signal Processing for Partial Discharge Analysis

Several signal processing techniques are used in PD-based predictive diagnosing tools for high voltage machines. One example of application is the design and implementation of conditioning circuits that shape amplitude and frequency of the measured signal for effective analog-to-digital conversion. This process comprises choosing the sampling rate accordingly, definition of binary representation of signal samples, and design the filtering stage to separate PD pulses from interference.

Design and implementation of filters is a vast area in signal processing with many technical choices to consider: digital or analog synthesis, linear or non-linear approaches, invariant or adaptive systems, just to name the main alternatives [13, 14]. Among the most common techniques applied to this subject [15], one can mention conventional filtering, wavelet-based denoising [16], adaptive noise cancellation [17], and artificial neural networks [18].

Another problem is PD analysis. Assuming a given observation time, it is verified that the number of PD pulses, associated with their peak amplitudes and point of occurrence in powerline cycle, provide distinct statistical patterns, from which one can identify the type of defect that generated the partial discharges. Such identification is important, as the risk for the high voltage machine changes according to the type of partial discharge. Phase-resolved partial discharge (PRPD) and pulse height analysis (PHA) diagrams constructed from statistics of recorded PD pulses have proved to be valuable tools for the estimation of both type and severity of partial discharges [19]. Intended to advance predictive diagnosing, recent approaches focus on either periodic or online PD monitoring. Artificial neural networks can be trained to automatically determine the PD type from features like PRPD or PHA. In general, training data are obtained from measurement campaigns in laboratory, where the experimenter can control the generation of PDs, for instance by setting the temperature or the abrasion condition at specific points of a stator bar. Once such patterns are learned by ANN classifiers, they favor the implementation of systems capable of continuous and automatic evaluation of insulation system condition [20].

1.6 Systems for Monitoring Partial Discharges

In 1989, the IEEE [21] defined a test to detect partial discharges (screening test) for bars and coils of the stator. With the operating frequency ranging from 8 to 100 MHz, the proposed strategy proved to be more sensitive to partial discharges caused by internal voids in dielectrics than the methods used until then. In [22], changes in the original scheme were proposed to guarantee repeatability of the experiments. Empirical evidence demonstrated that the lifetime of coils and bars can be extended by setting the operating voltage of such devices with the help of information obtained from the proposed test.

In [23], the insulation system of a 16 kV/250 MVA generator was investigated. Capacitive couplers, installed in the output bars of the generator, were employed in conjunction with a four channel fast sampling oscilloscope to register transient signals derived from partial discharges. Wavelets and other digital techniques were used for signal denoising. The measurements indicated the presence of small superficial discharges in all the phases and showed that the input impedance affects the monitoring sensitivity. In [24], the abrasion of the stator output bars is used to evaluate the insulation aging of the stator windings. The results show the correlation between aging time and two parameters: the skewness of phase resolved distribution and the high-frequency crest of the partial discharges.

In [25], some output bars, which have been operated for 7 and 22 years, were removed from generator to verify the state of their insulation systems. The authors performed a $\tan\delta$ evaluation and also investigated other parameters related to partial discharges, like the AC rupture voltage, PD interception voltage (PDIV), and maximum and total apparent electrical loads. All collected data were statistically analyzed. Results indicate the AC rupture and PD interception voltages, as well as the value of maximum electrical load, strongly depend on the bar service time: AC rupture and PD interception voltages reduces with the work-time and the maximum electrical load increases with time of operation.

Based on a large number of measured partial discharges, the authors in [26] proposed a system that separates the complete PRPD pattern, which is produced in general by the contribution of more than one generation mechanism, into sub-patterns that are due to only one type of partial discharge each. Such pattern separation approach improves the performance of classifiers, when applied to the identification of PD type in real settings.

Practical systems for monitoring partial discharges must face a challenging environment, where interference has many characteristics (white, narrow-band, and impulsive noise, and combinations of them) and partial discharge pulses are due to different mechanisms (delamination and interval void, for instance).

1.7 Summary

This book can serve as a practical guidance for students and engineers working in the field of high voltage machines and their maintenance. We believe that experienced researchers will also benefit from the book containing cutting-edge methods, practical examples and trends in this field. The book is written by a group of Brazilian specialists possessing a wide experience in electromagnetic modeling, numerical simulation, signal processing, machine learning techniques, electrical maintenance and online monitoring of PD in hydrogenerators. The book gathers the generally sparse knowledge required for practical work in hydroelectric plants into a single book.

The members of this group have been worked in this field about two decades. The information given in the book represents the collective experience and knowledge of the authors in PD research, measurement and analysis for hydroelectric power plants. A significant part of the presented results was obtained by the authors in their research projects and in field activities. As a basic background for reading this book we assume electrical engineering department programs of universities.

The book is organized as follows. Chapter 2 presents physics of PD, the main terms and concepts used for their description. Special attention will be given to PD that occur in stator insulation, presenting to the reader the sources of these discharges. Their characteristics, typical PD patterns in stator insulation and also sensors and methods to measure of PD are treated in Chap. 3. Chapter 4 addresses PD types in rotating machines, stator isolation faults, partial discharges measurement methods, the main characteristics of the PD measurement system and field measurement results. Chapter 5 is devoted to a resonance-based spectral method for pinpointing PD occurring in hydrogenerator coils. In Chap. 6, we discuss digital signal processing techniques applied to PD monitoring and classification. Chapter 7 contains some ideas of possible applications of ozone sensors for detecting and monitoring PD. Finally, some auxiliary material to different chapters is presented in Appendices.

References

1. Stone, G., Culbert, I., Boulter, E., & Dhirani, H. (2014). *Electrical insulation for rotating machines: design, evaluation, aging, testing, and repair* (2nd ed.). Wiley-IEEE Press.
2. Working Group A1.10. (2009). Survey of hydrogenerator failures. In *CIGRE - Conseil International Des Grands RÉseaux Électriques* (no. 392).
3. *Rotating Electrical Machines - Part 27-2: On-Line Partial Discharge Measurements on the Stator Winding Insulation of Rotating Electrical Machines. IEC/TS 60034-27-2*. International Electrotechnical Commission, Tech. Rep., 2012.
4. Stone, G. (2005). Partial discharge diagnostics and electrical equipment insulation condition assessment. *IEEE Transactions on Dielectrics and Electrical Insulation, 12*(5), 891–904.
5. Iris Power, *Iris Power website*, Retrieved January 24, 2023, from http://www.irispower.com
6. http://www.adwel.com. Accessed 24 January 2023.
7. Zhu, H., Green, V., Sasic, M., & Halliburton, S. (1999). Increased sensitivity of capacitive couplers for in-service PD measurement in rotating machines. *IEEE Transactions on Energy Conversion, 14*(4), 1184–1192.
8. Sedding, H., Campbell, S., Stone, G., & Klempner, G. (1991). A new sensor for detecting partial discharges in operating turbine generators. *IEEE Transactions on Energy Conversion, 6*(4), 700–706.
9. Stone, G., Sedding, H., & Costello, M. (1996). Application of partial discharge testing to motor and generator stator winding maintenance. *IEEE Transactions on Industry Applications, 32*(2), 459–464.
10. McDermid, W., & Bromley, J. (1999). Experience with directional couplers for partial discharge measurements on rotating machines in operation. *IEEE Transactions on Energy Conversion, 14*(2), 175–184.
11. Yee, K. (1966). Numerical solution of initial boundary value problems involving Maxwell's equations in isotropic media. *IEEE Transactions on Antennas and Propagation, 14*, pp. 302–307.

12. Taflove, A., & Hagness, S. C. (2005). *Computational Electrodynamics, The Finite-Difference Time-Domain Method* (3rd ed.). Artech House Inc.
13. Oppenheim, A., & Schafer, R. (2013). *Discrete-time signal processing: Pearson New International Edition*. Pearson Education.
14. Daryanani, G. (2009). *Principles of active network synthesis and design*. Wiley India Pvt. Limited.
15. Long, J., Wang, X., Zhou, W., Zhang, J., Dai, D., & Zhu, G. (2021). A comprehensive review of signal processing and machine learning technologies for UHF PD detection and diagnosis (I): Preprocessing and localization approaches. *IEEE Access*, *9*, 69 876–69 904.
16. Si, W., Qin, B., Li, Q., & Liu, H. (2019). A novel adaptive wavelet threshold estimation based on hybrid particle swarm optimization for partial discharge signal denoising. *Optik*, *181*, 175–184.
17. Lu, L., Zhou, K., Zhu, G., Chen, B., & Yana, X. (2021). Partial discharge signal denoising with recursive continuous S-shaped algorithm in cables. *IEEE Transactions on Dielectrics and Electrical Insulation*, *28*(5), 1802–1809.
18. Soltani, A. A., & El-Hag, A. (2021). A new radial basis function neural network-based method for denoising of partial discharge signals. *Measurement*, *172*, 108970.
19. Luo, Y., Li, Z., & Wang, H. (2017). A review of online partial discharge measurement of large generators. *Energies*, *10*(11), 1694.
20. Mantach, S., Ashraf, A., Janani, H., & Kordi, B. (2021). A convolutional neural network-based model for multi-source and single-source partial discharge pattern classification using only single-source training set. *Energies*, *14*(5), 1355.
21. IEEE recommended practice for voltage-endurance testing of form-wound bars and coils. (1989). In *IEEE Std 1043-1989* (pp. 1–18).
22. Mcdermid, W., & Bromley, J. (1999). Partial discharge screening test for internal voids and delaminations in stator coils and bars. *IEEE Transactions on Energy Conversion*, *14*(3), 292–297.
23. Birlasekaran, S. (2003). Identification of the type of partial discharges in an operating 16kV/250 MVA generator (pp. 559–562).
24. Yue, B., Chen, X., Cheng, Y., Song, J., & Xie, H. (2006). Diagnosis of stator winding insulation of large generator based on partial discharge measurement.' *IEEE Transactions on Energy Conversion*, *21*(2), 387–395.
25. Morin, R., Novak, J., Bartnikas, R., & Ross, R. (1995). Analysis of in-service aged stator bars. *IEEE Transactions on Energy Conversion*, *10*(4), 645–654.
26. Borghetto, J., Cavallini, A., Contin, A., Montanari, G. C., de Nigris, M., Pasini, G., & Passaglia, R. (2004). Partial discharge inference by an advanced system. Analysis of online measurements performed on hydrogenerator. *IEEE Transactions on Energy Conversion*, *19*(2), 333–339.

Chapter 2
Partial Discharges: Physics and Classification

2.1 General Characteristics of PDs

Partial discharges are electric discharges that occur between two or more electrodes in a medium, with energy levels lower than those encountered during full electric discharges. They are considered to be one of the earliest signs of electrical insulation failure. PDs are often found in electrical equipment such as power transformers, motors and generators, and they can cause significant damage to the equipment if left unchecked.

PDs can range in size and energy levels, often depending on the medium and the voltage of the system. High-voltage systems are more likely to exhibit larger PDs, and the energy of these discharges can range from a few millijoules up to several hundred joules. PDs can typically be divided into two categories: surface discharges, which occur on the surface of a material, and internal discharges, which occur inside the material. The amount of energy released during a PD is determined by the type of discharge and the intensity of the electric field. In general, PDs release small amounts of energy in comparison with the full discharges. The type of discharge can range from a spark, corona or arc to a low-level glow discharge. The shape of the discharge is also determined by the type of discharge and can range from a point source to a wide area of illumination.

PDs are rather difficult to detect and they are often unnoticed until they cause significant damage to the equipment. In order to detect PDs, a special device known as a partial discharge detector should be used. These detectors measure the changes in currents or electric field strength that are caused by the PDs and they can be used to pinpoint the location of the discharge. PDs can be caused by a variety of factors such as contamination of the insulation, poor insulation design and electrical stress. Contamination of the insulation can lead to a breakdown of the insulation which can provoke PDs. Poor insulation design can cause the electric field to become too strong and can cause PDs as well.

V. Dmitriev et al., *Partial Discharges in Hydroelectric Generators*, Power Systems,
https://doi.org/10.1007/978-3-031-36604-8_2

The size and shape of a PD is determined by the type of material in which it is occurring. In general, PDs tend to have a small size, but this can vary depending on the material and the intensity of the electric field. PDs also tend to have a short duration, typically lasting only a few microseconds. The location of the PD is determined by the geometry of the material in which it is occurring. PDs on the surface of the material tend to be more localized and intense, while PDs within the material tend to be more diffuse. The rate of spread of a PD is determined by the electrical properties of the material in which it is occurring. PDs tend to move through the material with a speed which can vary depending on the specific material. Additionally, the rate of spread can be influenced by the presence of conducting particles or defects in the material.

PDs create electrical noise in the electrical system which can cause interference in the system. PDs can provoke overheating of the equipment which can lead to further damage if not addressed. PDs can also cause insulation breakdown and complete electrical breakdown of the system.

In order to properly detect and monitor PDs, it is important to use a partial discharge detector. These detectors can provide accurate and reliable data about the environmental conditions and the electrical system, which can help to prevent further damage to the equipment. Proper monitoring and detection of PDs can help to reduce the damage caused by the discharges and can help to ensure the safety and reliability of the electrical system.

The PDs can be classified into two types: continuous discharges and intermittent discharges. The continuous discharges occur when the electric field strength is greater than the breakdown value of the material, and the current is continuously conducted through the gas or vapor produced by the ionization process. The intermittent discharges occur when the electric field strength is smaller than the breakdown value of the material, and the current is conducted intermittently through the gas or vapor produced by the ionization process.

The characteristics of the PDs are affected by the characteristics of the material, the electric field strength, the temperature, the pressure and the humidity. The characteristics of the PDs can be used to determine the condition of the insulation system and to detect defects that can be in the insulation system. It is important to understand the physical characteristics of PDs in order to properly assess the risk of damage to the material. This includes understanding the amount of energy released, the type and shape of the discharge, the location of the discharge and the rate of its spread.

2.2 Physics of PDs

2.2.1 Ionization of Insulator Materials

PDs occur in electrical insulation systems when the electric field strength exceeds the breakdown threshold of the material, leading to the conduction of current through the gas or vapor produced by the process of ionization. Thus, the breakdown of the material occurs when the electric field intensity is greater than the dielectric strength of the material.

Ionization of the material is the process of separating the electrons of the molecules of the insulation material, producing a gas or vapor of ions and electrons. This process is more easily observed in gases and liquids where the ionization process occurs more quickly. In solid materials, the ionization is more difficult to observe as the electrons are more tightly bound to the atoms. The breakdown of the material is a stochastic process, meaning it cannot be accurately predicted. It is possible to estimate the time of occurrence of the breakdown by taking into account the electric field strength and the characteristics of the material. However, statistics during breakdown can be assessed.

The ionization of insulator materials in hydrogenerators is an important aspect of the safety and operation of hydroelectric power plants. It is a process in which ions from the environment are absorbed into the insulating material, which can lead to a decrease in its dielectric strength. This can provoke formation of electrical arcs, which can cause a variety of problems, including breakdown of the insulator material and even fires and explosions. The ionization of insulators is a complex process and it is affected by a number of factors. They include the type of insulating material, the operating environment and the presence of certain contaminants in the air or water. It is also affected by the humidity of the air and the temperature and the electrical current passing through the insulator.

The ionization of dielectrics can be prevented by the use of appropriate materials and designs. The use of materials with high dielectric strength, such as polyethylene, and the use of designs that reduce the exposure of the insulating material to contaminants are important steps in preventing the problems caused by ionization.

The most common method used to detect the ionization of insulators in hydrogenerators bars is the use of a high voltage test. This test is used to measure the breakdown voltage of the insulator material, which can indicate if there is a problem with the ionization strength of the material. The design of hydroelectric power plants must also include measures to protect against the ionization of insulators in their generators.

2.2.2 *Characteristics of PDs*

According to [1], to understand the phenomenon of PDs, it is necessary to know how the atoms or molecules of dielectric material are ionized. For practical cases of simple ionization, the relation between the potential difference V and the electric field $\vec{E}$ is [2]

$$\nabla V = -\vec{E}, \tag{2.1}$$

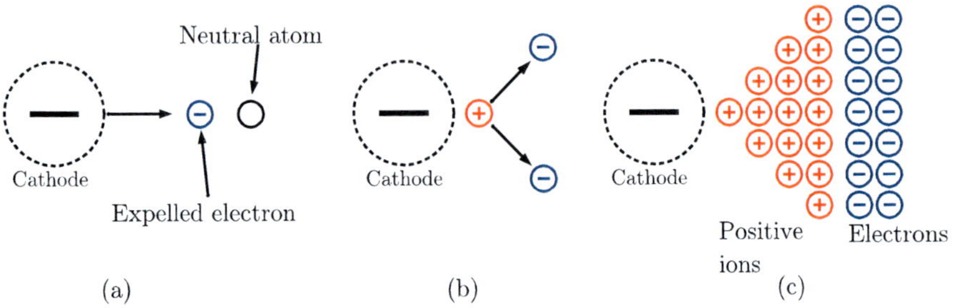

Fig. 2.1 Electron avalanche starting from a negative electrode: (**a**) beginning of ionization process; (**b**) release of a pair of electrons: the impact of an electron with a neutral atom releases one additional electron and leaves a positive ion behind; (**c**) multiplication: electrons move producing positive ions as more electrons are released

where ∇ is the operator gradient and the dielectric is considered to be linear and isotropic. As described in [1], the initial electron, which has lost most of its speed in the collision, and the electron ejected from the air molecule which also has a low velocity, are accelerated by the electric field. In the next collision, each electron is capable of ionizing one molecule of air. After the second collision, there are four electrons capable of ionizing other atoms, and so on, with the number of electrons doubling at each collision. This process is known as the avalanche of electrons, always being initiated with a single free electron exposed to an intense electric field. Figure 2.1 illustrates the sequence of ionization of the atoms, i.e. process of the avalanche of electrons starting from negative electrode. In Fig. 2.1a, the process begins with a single electron. When it collides with a neutral atom, a pair of electrons is created Fig. 2.1b and, in the process of multiplication Fig. 2.1c, electrons move and create positive ions, thus multiplying their number.

The positive ions left behind in the avalanche of electrons move towards the negative electrode. However, due to their mass, which is approximately fifty thousand times larger than that of the electron, they move at a low velocity. Having a positive charge, these ions attract free electrons and, when one of them is captured, another neutral molecule of air is formed. The energy level of a neutral molecule is lower than that of a corresponding positive ion. Therefore, when a free electron is captured, the molecule emits an energy quantum. This energy quantum is equal in magnitude to the energy initially required to dislocate the initial electron from its molecule. An electromagnetic wave is radiated and, to molecules of air such as oxygen and nitrogen, this radiation falls within the spectrum of visible light [1, 3]. When the energy source is depleted, the ionization process will stop, but the recombination will continue until there are no more free electrons or positive ions.

The electrons and ions generated by this process are capable of conducting current between the electrodes and dissipating a significant portion of the source power, resulting in a spark between the electrodes. When this sparking occurs, it is said that the dielectric material has been ruptured. Since the spark usually does not traverse the entire distance between the electrodes, it is referred to as a partial discharge [3].

The avalanche of electrons illustrated in Fig. 2.1 has a certain number of electrons per second, which can vary from hundreds to 10^{22} electrons per second in a typical period of 100 nanoseconds. To quantify the process, Coulomb is used as a unit, which is equivalent to the charge of 6.2×10^{18} electrons. As one Ampere is defined as one charge flow of one Coulomb per second, the current of the avalanche of electrons can vary from 10^{-17} A to thousands of Amperes [3].

2.2.3 *Propagation of Pulse in the Stator Windings*

According to [4], the pulse of a PD has an extremely fast rise time and a short width. The oscillation period, rise time and magnitudes of subsequent peaks vary from pulse to pulse. These characteristics usually depend on the machine geometry, the pulse location and the insulation material. The measured PD signals will have distinct characteristics depending on the properties of each winding, as well as the location and type of the PD source. Moreover, PDs are subject to attenuation as they propagate through the winding and, depending on their location, can become undetectable by the measurement circuit. The next figure, adapted from [5], shows how the PD pulse is attenuated with distance and its rise time decreases as the pulse propagates towards the measurement circuit (Fig. 2.2).

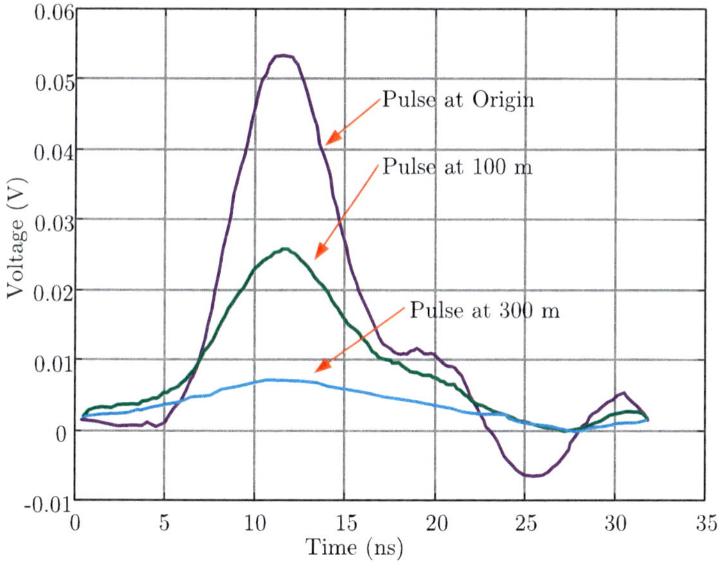

Fig. 2.2 Temporal distortion of PD pulse acquired along the measurement circuit. Source: adapted from [5]

Figure 2.2 shows the distortion of the pulse of the PD along the measurement circuit. The pulse is generated by the PD and, as it propagates through the circuit, it is altered due to the effect of the circuit components, such as capacitors and inductors. The shape of the pulse is typical for PDs and is used for their recognition and classification. The amplitude of the pulse varies along the circuit and can be more or less distorted depending on the type of circuit used for the measurement.

The distortion of the PD pulse along the measurement circuit is an important factor in the PD detection process, as it affects the accuracy of the PD location and classification. Therefore, it is important to understand the behavior of the PD pulse along the circuit and to design the measurement circuit accordingly. The pulse distortion also affects the signal processing techniques used to identify the PDs, as the distortion of the pulse can lead to false detection.

In order to analyse the pulses with high frequency components in its spectrum, the stator winding can be modeled as a circuit of distributed parameters, a transmission line where the PD pulses propagate from their origin to the measurement circuit, as shown in the simplified scheme of Fig. 2.3. From the place of origin to the measurement circuit, the PD pulse propagates through a network of distributed parameters, composed of serial inductances, parallel capacitances and shunt capacitances to the ground.

Due to phenomena of attenuation, distortion and coupling between phases of the generator, the amplitude and shape of the PD waves captured by the measurement circuit differ significantly from those that occurred at the point where the PD originated.

The PDs can be observed in the electrical network by performing a measurement of the PD current in the primary windings of the machine. The PD current can be measured by connecting a measuring device to the primary windings, such as a current probe or a differential transformer [6]. The transfer function between the partial discharge source and the measurement circuit is thus unknown, and depends on the specific project of each machine, which determines the frequency response of the stator winding. Therefore, the energy in the PD origin cannot be measured directly. This fact prevents calibration of PD measurements in synchronous machines. The measurement in synchronous machines usually registers the amplitude of PD in

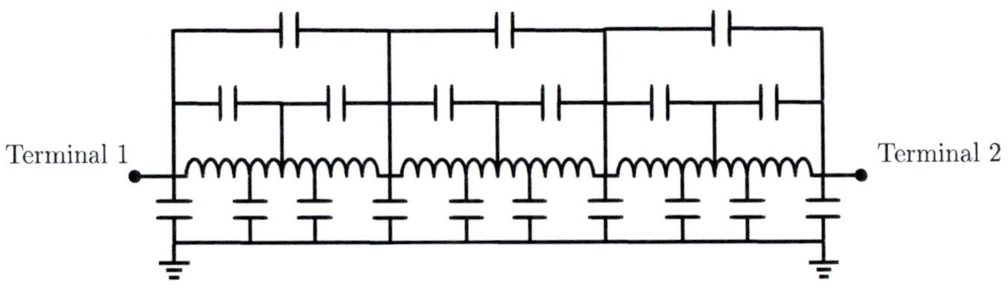

Fig. 2.3 Schematic representation of a network of inductances and capacitances modeling the stator windings. Source: adapted from [6]

mV, rather than in pC as recommended by technical specification IEC 60270 [7] for insulations modeled with concentrated parameters.

Since a PD is a localized electric discharge in a dielectric that do not completely bridge the electrodes, like in an arc, detecting PDs is important because they are associated with electrical insulation degradation and are considered a precursor of insulation failure. PDs can be classified according to the type of environment in which the discharge occurs, for example, air, liquid or solid. PDs in air are caused by the presence of impurities, such as dust particles or air bubbles. PDs in liquids occur when the dielectric is contaminated by water or other contaminants, or when a high electric field is applied. PDs in solids are often caused by the presence of defects such as cracks in the dielectric material.

2.3 Features of PDs and Their Measurements

PDs have various distinctive characteristics, such as frequency, amplitude, duration and phase. The frequency of PDs depends on the type of environment in which the discharge occurs, and on the type of impurity or defect present. The amplitude and duration of PDs are determined by the size of the impurity or defect and by the electric field strength. Finally, the phase of PDs depends on the type of environment in which the discharge occurs and on the type of impurity or defect present.

PDs can be detected and measured using various techniques, such as electrical measurements, optical measurements and acoustic measurements. Electrical measurements involve measuring the current, voltage, or power of the discharge. Optical measurements involve measuring the light emitted by the discharge, while acoustic measurements involve measuring the sounds generated by the discharge.

Some definitions of magnitudes related to PD measurement, according to international standards [4, 8], and [7], will be presented. Let Ω be a PD activity. It is defined as the number of PD pulses per unit time [8]. The partial discharge magnitude Δ is the ratio between the maximum value and the minimum value of the associated voltage impulse, and is expressed in picocoulombs (pC). The partial discharge frequency f is the number of discharges per unit of time, and is measured in Hertz (Hz). The PD phase angle θ is the phase difference between the voltage and the current of the associated PD impulse, and is measured in degrees. The PD inception voltage U_i is the voltage at which the PD activity begins. The PD extinction voltage U_e is the voltage at which the PD activity ceases [4]. Finally, the PD power P is the ratio between the energy transferred by the PD and the time it takes to transfer the energy and is expressed in watts (W) [7]. In summary, PDs are characterized by their activity, magnitude, frequency, phase angle, inception voltage, extinction voltage and power.

2.3.1 Magnitude of Partial Discharge With Greater Repeatability Q_m

The magnitude of PD with greater repeatability Q_m (also referred to as the apparent charge) is an important parameter which is used to characterize the PDs. It is defined as the total charge Q_t of a single PD pulse divided by the number of pulses during the observation interval N. It is expressed by

$$Q_m = \frac{Q_t}{N}. \tag{2.2}$$

The magnitude of PD with greater repeatability can be measured in several ways. The most common method is to measure the peak-to-peak voltage of PD pulses and multiply this value by the capacitance of the discharge circuit. The magnitude of PD with greater repeatability is an important parameter for PD characterization and it is a useful parameter for predicting PD behavior in different conditions. It is important to note that the magnitude of PD with greater repeatability is a very sensitive parameter, which is affected by several factors, such as the discharge source, the type of insulation material, the temperature and the atmosphere. Q_m [9] is defined as a magnitude registered by a measurement system that has the response to a pulse train, according to the International Electrotechnical Commission (IEC) 60270 standard [7]. The magnitude is associated with a pulse repetition rate of a certain number of pulses per second, with 10 pulses per second being the recommended rate.

The magnitude of PDs can be quantified by the charging current i_c in picoamperes (pA), the magnitude of the induced charge Q_M in nanocoulomb (nC), by the magnitude of the field strength E_m in kilovolts per meter (kV/m) or by a corresponding voltage (mV).

2.3.2 Normalized Quantity Number NQN

Normalized quantity number (NQN) is a parameter used to classify PDs according to the PD magnitude. It is defined as the ratio between the root mean square (RMS) voltage measured in the discharge channel and the RMS voltage of the applied voltage waveform, that is

$$NQN = \frac{V_{rms}}{V_{rms}^{peak}}. \tag{2.3}$$

NQN is a dimensionless quantity and it is usually normalized to $1kV_{peak}$. Therefore, a NQN of 0.2 indicates that the PD magnitude is $0.2kV_{peak}$. NQN is suitable for

PD magnitude classification since it is not dependent on the waveform parameters (frequency, rise time, etc.). According to IEEE 1434-2014, NQN is defined as the area determined by under the line formed by the number of pulses (logarithmic scale) and amplitude (linear scale). We may thus write

$$NQN = \frac{1}{T_r} \sum_{k=1}^{N} \frac{Q_k}{Q_m}, \tag{2.4}$$

where T_r is the repetition time, N is the number of pulses, Q_k is the charge of the k-th pulse and Q_m is the maximum charge of the discharge. The number of pulses per second (NPS) can be expressed in terms of a logarithmic scale, denoted by Π. It is defined as $\Pi = 10^{\log_{10}(n_p)}$, where n_p is the number of pulses per second. This allows us to compare the intensity of different PDs, regardless of the measurement system used.

The gain G of a PD measurer is defined as the ratio between the charge collected by the PD measurer and the charge created by the discharge. G typically ranges from 10^3 to 10^6 and depends on the capacitance C_{meas} of the measurer and the impedance Z of the measurer-ground system, according to

$$G = \frac{C_{meas}}{Z}. \tag{2.5}$$

G is an important parameter that should be carefully set for performing accurate measurements.

The measurement units used for PD tests can be pC (picocoulombs) or mV (millivolts) and this depends on the test configuration used. The pC unit represents the amount of electrical charge released by the partial discharge, while mV represents the voltage generated by PD. The group that follows the IEC 60270-2000 standard [7] recommends the use of the pC as the measurement unit. Conversely, the group led by Americans and Canadians, which grounds itself on the recommendations of the IEEE guide, uses the magnitude represented by mV. The major difference between these two methodologies lies in the calibration of the measurement circuit. While the pC measurement method requires calibration, the mV measurement method does not. Calibration for pC measurements allows for precise and normative quantification of PD amplitudes. This allows for a quick and simple evaluation of a certain equipment, either new or used. Moreover, it provides a possibility of comparison between similar equipment, such as current transformers (CTs). For example, CTs must present levels below 10 pC when tested in laboratories with sufficient quality to measure amplitudes of this order.

The phenomenon of PDs is, by nature, difficult to detect. Without access to the physical discharge within the dielectric material, the measurement is limited to recording oscillations of a few mV over an operating voltage of the order of kV. As a circuit of distributed parameters, the synchronous machine insulation cannot be evaluated in the same way as devices with concentrated parameters, as suggested by the standard IEC 60270-2000 [7]. The indeterminacy of the location where the discharge occurs in the stator insulation implies indeterminacy of the transfer function between its origin and the measurement circuit. This makes it impossible to quantify the attenuation and the distortion that signals suffer due to their propagation through the distributed circuit of the machine winding.

The pulse repetition rate (PRR) n is a parameter that describes the PD behavior in a certain insulation system. The PRR is a measure of the number of PDs that occur in a unit of time, and is expressed as

$$n = N_1/T_\mathrm{i}, \tag{2.6}$$

where N_1 is the total number of PDs in the time interval T_i. PRR is a useful parameter for characterizing the PDs since it provides information about the frequency of PDs that occur in an insulation system. Furthermore, it is possible to estimate the severity of the PD defect by measuring its PRR. The higher the PRR, the more severe the defect is. PRR can be used to compare different insulation systems with different materials or geometries. It can also be used to compare different locations within the same insulation system. For example, the PRR can be used to compare the PDs at the same location of different insulation systems, or to compare the PDs at the same location over time. It can also be used to identify the source of the PDs. By measuring the PRR at different locations of the insulation system, it is possible to identify which location the source of the PDs has. The average number of pulses of PDs per second, measured over a chosen time interval, is a parameter for diagnosing insulation degradation. In practice, only pulses that exceed a specific intensity threshold are typically taken into account.

The phase angle of the PDs is an important parameter for the assessment of electrical insulation condition, associated to the pulse occurrence time (POT). The POT is the time elapsed between two pulses and is measured in microseconds. The PD phase angle is related to the time of occurrence of the discharge in relation to the applied voltage. The POT and phase angle depend on the PD process, which is subcategorized in four classes: corona, streamer, spark and combination [8]. The PD phase angle is usually measured in degrees (°) and its values range from 0° to 360°. It is important to note that the PD phase angle is affected by the POT, and its values can vary significantly depending on the PD type. PD phase, in degrees, can be calculated by

$$\varphi_i = 360\frac{t_p}{T}, \tag{2.7}$$

where t_p is the instant of partial discharge initiation (or peak, depending on the system) and T is the sinusoidal period.

2.3.3 Inception Voltage of PDs

The inception voltage (U.I.N.) of PDs is the minimum voltage on an insulation system that can produce the PD phenomenon. This voltage is a function of the type of discharge, the type and condition of insulation, the geometric configuration of the system, the environmental conditions, and the type of applied voltage [7]. This voltage can be measured in a laboratory using the Coulomb method, which consists of applying a periodic voltage on the insulation system and measuring the charge of the PD in a certain period of time.

The inception voltage of the PD has a very important role in electrical insulation systems, as it is a parameter that can be used to evaluate the quality of insulation and to compare different materials or configurations. Lower voltages at which PDs are observed, when the voltage applied to the object under test is gradually increased from a level at which no PDs are observed.

The inception voltage of PDs in hydrogenerators is largely dependent on the type of insulation used. Generally speaking, insulations with high breakdown strength can withstand higher partial discharge voltages before breakdown occurs. This is due to the fact that high breakdown strength materials are more resistant to electrical currents, thus allowing higher voltages to be applied before a breakdown occurs.

The environment in which the hydrogenerator is operated can also have an effect on the inception voltage of PDs. In particular, humid air can reduce the breakdown strength of insulations, thus resulting in a lower inception voltage. This is due to water molecules in the air being able to penetrate the insulation, resulting in a lower breakdown strength.

In addition to the type of insulation used and the environment in which the hydrogenerator is operated, the size of the generator can also have an effect on the inception voltage of PDs. Larger generators typically have larger internal components, resulting in higher voltages being required to cause a breakdown.

The operating frequency of the hydrogenerator can also have an effect on the inception voltage of PDs. Generally speaking, higher frequencies can result in a higher inception voltage due to the fact that the high frequency causes an increase in the electric field strength of the insulation.

In general, theinception voltage is obtained experimentally from the characterization of PD activity as a function of the applied test voltage. This process is tedious and time consuming. For this reason, several PD extinction voltage models have been proposed to estimate the PD extinction voltage from measured or calculated parameters of insulation systems.

The inception voltage of PDs in hydrogenerators can vary significantly depending on the various factors discussed above. Therefore, it is important to take all of these factors into account when assessing the safety and reliability of hydrogenerators. By understanding the inception voltage of PDs, hydrogenerators can be properly maintained and operated safely.

2.3.4 Extinction Voltage of PDs

The extinction voltage of PD is defined as the voltage at which PD activity ceases. In power systems, the PD extinction voltage is a function of the insulation structure and is related to the dielectric strength of the insulation.

In general, the extinction voltage is obtained experimentally by characterizing PD activity as a function of the applied test voltage. Several models have been proposed to estimate the PD extinction voltage from either measured or calculated parameters of insulation systems.

The upper voltage at which partial discharges (PDs) cease when the voltage is gradually decreased from an upper value to the starting voltage of the discharge is typically used in practice to determine the starting and extinction voltages. Specifically, only PDs with an intensity greater than a specified low value are considered for this purpose.

2.3.5 Phase-Resolved PD Analysis

The PDs are usually analyzed using a special type of instruments called phase-resolved partial discharge (PRPD) analyzer. The analyzer is usually connected to the electrical system using a special type of probe. The probe is used to measure the electric field generated by the discharge and sends the data to the PRPD analyzer, which then calculates the amplitude and phase angle of the discharges.

In order to accurately measure the PDs, it is important to select the correct type of PRPD probe for the application. The type of probe depends on the type of electrical system being monitored, as well as the type of insulation material of the system. The probe must be properly calibrated and positioned in order to obtain accurate measurements. PRPDs are important metrics in the evaluation of the reliability of an electrical system. By accurately measuring the amplitude and phase angle of PDs, engineers can better monitor and diagnose the insulation state of electrical equipment. The selection of the correct type of PRPD probe, as well as the proper calibration and positioning, are key factors in obtaining accurate measurements.

According to [2], the analysis of association to PDs phase is currently considered one of the most effective methodologies for identifying PD sources. This method has been incorporated into the most modern instruments for measuring PDs. In the analysis, it is initially assumed that the test voltage or system voltage is constant and the phase angles of this wave are discretized into ranges. The PD measurement instrument quantifies the PD amplitudes related to the phase angle φ_i of the test voltage or system voltage during a certain time interval. During this time interval, the amplitude values are accumulated and can be expressed as the pulse repetition rate n.

Figure 2.4 shows the PDs signals of a real generator. This generator was monitored by a PRPD system. The PRPD system can detect PDs in a very wide

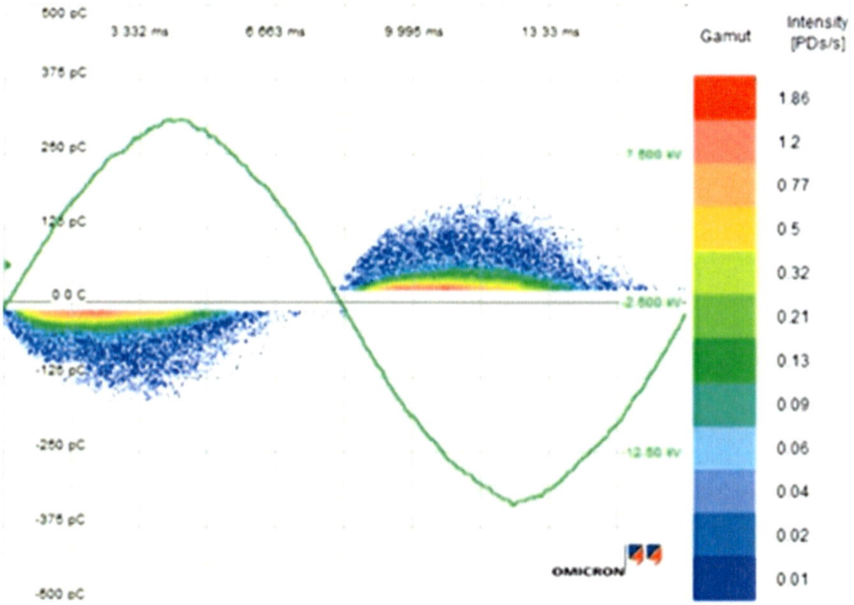

Fig. 2.4 An example of PRPD map

frequency range, from kHz to GHz. Table 2.1 presents some PRPD patterns obtained in the measurements of hydrogenerator stator bars in laboratory. These patterns are compatible to standards for different types of PDs, which can be used to diagnose the system health. In particular, the PRPD patterns can be used to identify the sources of PD, such as internal voids, slot discharges, surface discharges and delamination.

2.3.6 *Measurement of PD Signals*

Given that PDs involve a flow of electrons and ions over a very small distance in a short period of time, a small electrical current flows along the machine winding every time when a PD occurs. The total current is determined by the transportation of a certain amount of charge. This current flow creates a detectable electric voltage. The voltage measured at a point of the winding (with respect to ground) is proportional to the current flow, as well as the impedance of the winding. The latter is a function of the frequency and the inductance of the winding.

One of the methods for detecting PDs is measuring the voltage pulse that accompanies them or the resulting current pulse. These quantities can be measured in circuits far from the discharge site. It is worth noting that in the winding or bar of a hydrogenerator, hundreds of discharges can occur per second, resulting in hundreds of electrical pulses.

Table 2.1 Typical PD patterns: phase-resolved partial discharges (PRPDs)

Discharge type	PRPD
Internal voids	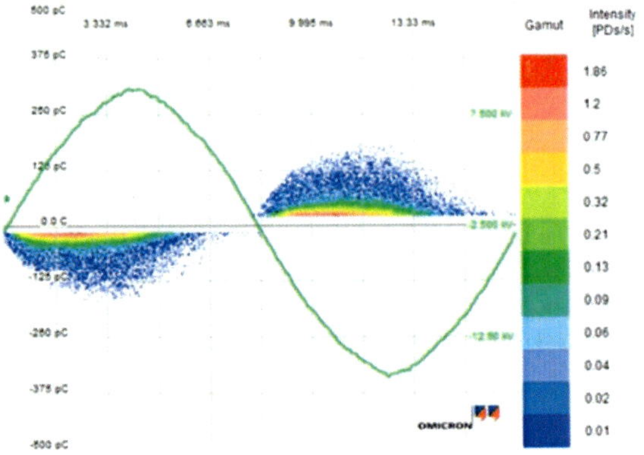
Slot discharge	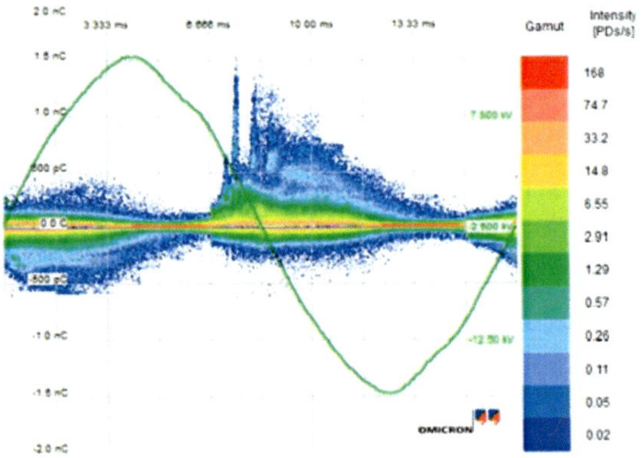
Surface discharge	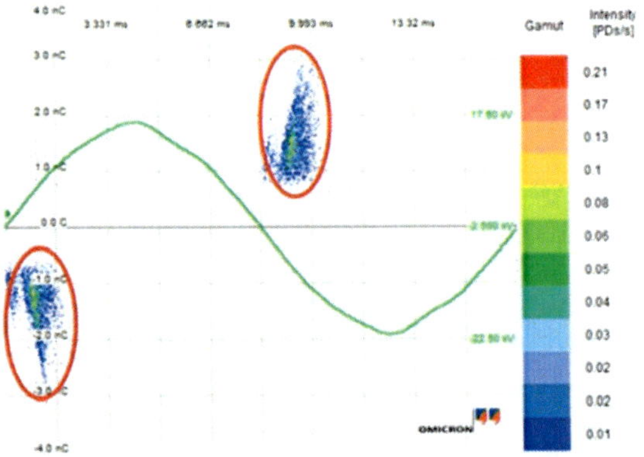

(continued)

Table 2.1 (continued)

Discharge type	PRPD
Delamination	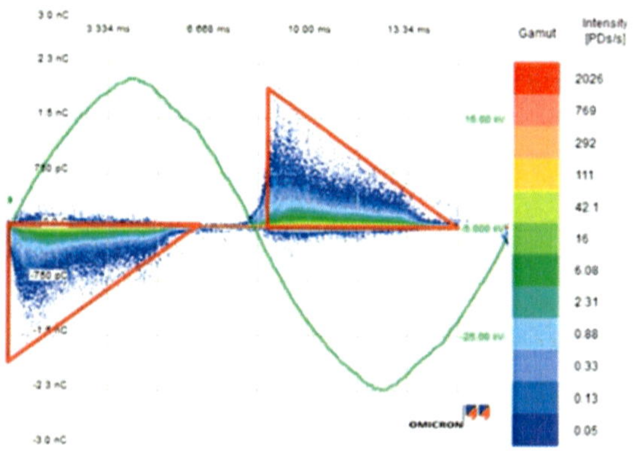

The high voltage coupling capacitor, usually connected to the terminal phase or winding, can detect voltage pulses. Its high impedance at industrial frequencies, combined with its low impedance to high-frequency voltage pulses of the PDs, makes it an ideal device for PD detection.

The output of the couplers are voltage pulses that can be measured using an oscilloscope, spectrum analyzer or PD measuring device. To determine the characteristics of PDs, the current and voltage pulses must be measured simultaneously. This can be accomplished by connecting a current transformer and voltage coupler in series. The current transformer measures the current flow of the PD event while the voltage coupler measures the voltage. The acquired data can then be used to calculate the magnitude and duration of the PD event.

2.3.7 Capacitive Coupling

Capacitive coupling is the most common method for detecting PDs. It is mainly used for online monitoring purposes. This is because PDs are detected by changes in the capacitance of the insulation, which is monitored by a capacitor (coupling capacitor) connected in parallel with the insulation [10]. Capacitive coupling is usually performed by a coupling capacitor with a capacitance of 1 to 10 nF, followed by a filter and a current amplifier. According to [7], the coupling capacitor serves to transfer the high-frequency spectrum of winding PD signals to the coupling device and also to mitigate the system voltage for low frequencies.

The value of capacitance of the coupling capacitor should be chosen considering both the capacitance of the object being evaluated and the desired frequency range for the measurements. In fact, when connected to the stator winding, the coupling capacitor functions as a high-pass filter with a termination resistor that can range from 500 up to 2000 Ω. Commercially, some standard values of coupling capacitors have been established. Therefore, it is possible to find capacitors of 80, 220, 550 and up to 1000 pF. As an example, a typical configuration in the Brazilian electrical system is to use a 80 pF coupler with a resistance of 690 Ω.

The installation and number of couplers must be carefully analyzed and debated, both from technical and economic perspectives. The number of couplers is related to the machine's physical size, as previously mentioned, since high-frequency signals suffer strong attenuation in the stator windings. Therefore, if it is desired to increase the area of the winding being analyzed, a larger number of couplers must be installed. However, in most cases for machines of small to medium size, one coupler per phase is enough for a good evaluation of the insulation state. Figure 2.5 shows the coupling in the interior of the winding and Fig. 2.6 presents the coupling at the exit of the phase.

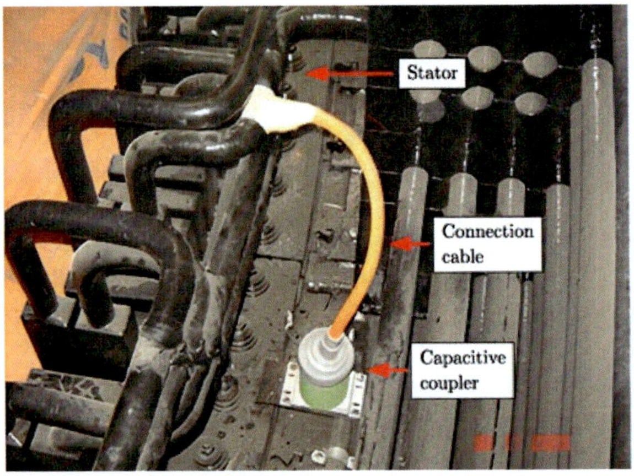

Fig. 2.5 Capacitive couplers installed in generator winding

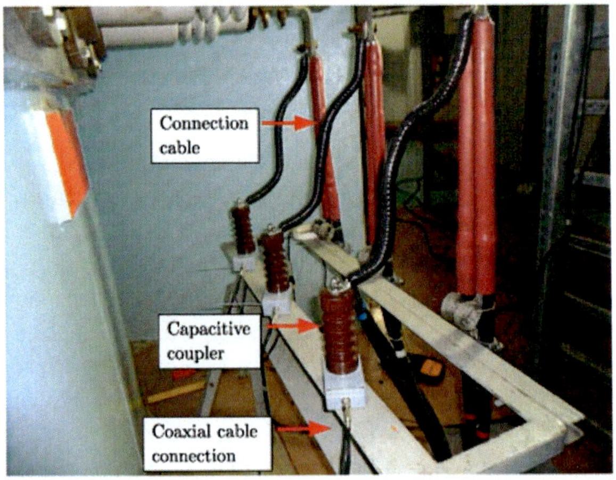

Fig. 2.6 Capacitive couplers installed in a phase output

The capacitive coupler is composed of two capacitors connected in series by a coaxial cable. The first capacitor is connected to the winding insulation, and the second one is connected to ground. In this configuration, PDs are characterized by a voltage rise at the output of the device, which is then used to detect them.

The couplers are composed of a capacitor connected in series with a resistor, as seen in Fig. 2.6. The capacitor allows the detection of the electrical pulses generated by the partial discharge, while the resistor limits the maximum current, thus avoiding arcing or other damages to the device. This combination of a capacitor and a resistor enables the capacitive coupler to detect PDs in a safe and reliable way.

Associated to the coupling capacitor is the measurement impedance which can range from a simple commercial resistance to a circuit comprising capacitors, inductors, and resistors. The impedance measurement is illustrated by Fig. 2.7. The purpose of this circuit is, in addition to capturing PD signals, to provide a voltage reference signal at the nominal frequency. The concept is to relate the parameter to its characteristic frequency. At high frequencies, the inductor L_m acts as an open circuit, while the capacitor C_m acts as a short circuit; and, conversely, at low frequencies, the capacitor acts as an open circuit and the inductor as a short circuit. Therefore, the capacitor C_m in series with the parallelly-connected resistor R_m and inductor L_m form a voltage-dividing system at 60 Hz frequency, where the inductor acts as a short circuit. At higher frequencies, it is an open circuit and only the resistance is useful in this case, allowing for PD signals to be detected.

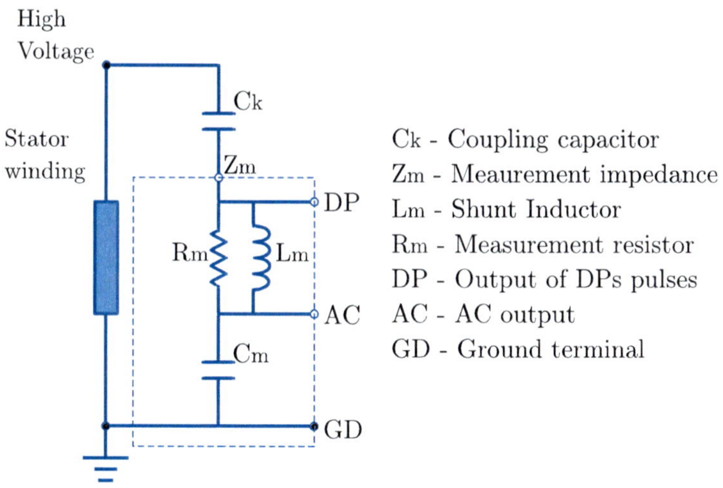

Fig. 2.7 PD measurement impedance. Source: adapted from [7]

2.4 Types of PDs in the Stator Windings

The purpose of this section is to provide an insight into the typical patterns of PDs that can occur in stator insulation. The various characteristics of different PD sources found within stator insulation are important for the successful application of various diagnostic methods. Depending on the nature of the medium in which they occur, PDs can be classified as:

- **Surface Discharges**: on the surfaces of solid dielectrics;
- **Internal Discharges**: within solid dielectrics;
- **Gaseous Discharges**: in gases.

PDs occurring in the windings of the stator can be further classified as:

- **Conductive Discharges**: occur when the insulation between the wires of the stator windings is compromised;
- **Capacitive Discharges**: emerge when the insulation between the layers of the stator windings is compromised.

In accordance with [8], the sources of PDs in hydrogenerators are classified into four groups: internal discharges, slot discharges, discharges in the end-winding, and discharges caused by conductive particles. Internal voids in insulating materials cause a reduction of the dielectric strength of the material and these voids can be the origin of PDs. The presence of voids can be caused by the use of materials with different thermal expansion coefficients and the consequent formation of cracks during temperature changes. In addition, the presence of voids can be caused by the presence of impurities, such as air bubbles and cavities in the material used for insulation. Internal voids can also be caused by the presence of moisture inside the insulating material. The presence of moisture causes the formation of micro-voids due to the process of molecular diffusion, which can lead to the initiation of PDs in the material. Internal voids can be detected by analyzing the electrical behavior of the material. Measurement of the partial discharge current and the dielectric strength of the material can be used to analyze the presence of voids. In addition, the presence of voids can be detected by performing a physical analysis of the material, such as measuring the thickness of the material or observing the surface of the material under a microscope. Internal voids can be avoided by using materials with similar thermal expansion coefficients, properly drying the material before use, and using materials with low impurity content.

Although manufacturing processes attempt to minimize internal voids, they are not eliminated. For instance, the system of insulation composed of mica leaves and fiberglass fabric impregnated with a synthetic resin is commonly used in high voltage rotating machines. The mica present in the insulation system acts to impede the growth of PDs, thus preventing the complete rupture of the dielectric. As long as the voids remain small and do not grow significantly over time, the operational reliability of the system is not compromised.

2.4.1 Internal Delamination

Internal delamination consists of separation of the insulation layers of a dielectric material. When PDs occur inside the insulation material, they can cause degradation of the insulation and consequently reduce the electrical strength of the material. Internal delamination is usually caused by mechanical stress, such as compression or tension, which can lead to the formation of pockets of air between the insulation layers. This separation of the layers creates a path of lower electrical resistance, leading to the accumulation of electrical charges and the formation of PDs.

The most common method of detecting internal delamination is ultrasound imaging. This technique utilizes ultrasound waves to measure the sound speed of the material and the attenuation of the waves. The sound speed is used to measure the homogeneity of the material and the attenuation is used to detect the presence of air bubbles between layers.

The internal delamination in the stator insulation can be caused during the manufacturing process or by the coils overheating. When the insulation system is continuously exposed to this overheating, the organic resins tend to lose their mechanical rigidity, leading to the dislocation of the insulation layers and resulting in a phenomenon known as layer delamination. When the delamination of insulation layers occurs, the internal copper conductors remain exposed and are prone to vibrate, resulting in high-energy PDs that can significantly compromise the insulation.

This kind of failure can lead to PDs in the form of sparks, arcs or streams of ions [11] which can cause severe degradation of the insulation. As previously stated, delamination can also be identified by visual inspection or by non-destructive techniques, such as infrared thermography and ultrasound methods. However, electrical measurements can be used for on-line monitoring of the condition of the insulation.

2.4.2 Delamination Between Conductors and Insulation

PDs on hydrogenerators can cause significant damage to the equipment if not addressed in a timely manner. One particular phenomenon that can occur is delamination between conductors and insulation. This is caused by the formation of an electrical breakdown of the insulation material due to the accumulation of PDs, resulting in the formation of a void between the conductor and the insulation. This is a significant issue for hydrogenerators as it can reduce the performance and reliability of the equipment.

In order to identify and diagnose partial discharge-induced delamination, specialized monitoring techniques are used. These techniques measure the electrical properties of the insulation material such as the insulation resistance, dielectric strength and capacitance. These measurements can be used to determine if there

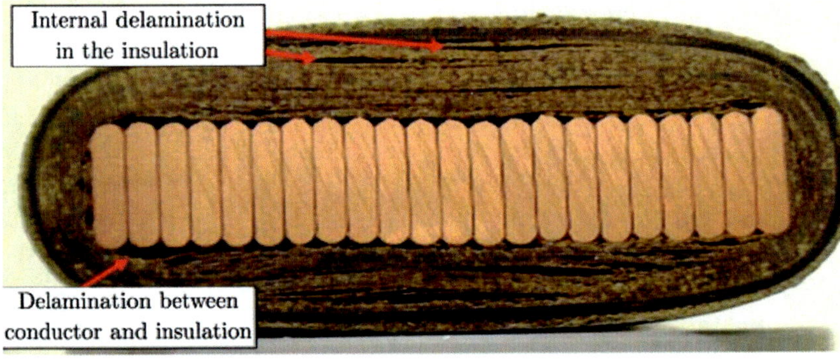

Fig. 2.8 Transverse section of a stator bar, where one can observe internal delaminations in the insulation

is an accumulation of PDs which can lead to delamination. Additionally, visual inspections can be used to identify signs of delamination.

The most effective way to address delamination is the application of appropriate preventive maintenance procedures. This includes regular inspections of the insulation material, as well as testing with specialized monitoring equipment. Any damaged insulation should be replaced immediately to prevent further damage.

In addition to preventive maintenance, there are a number of other approaches that can be taken to reduce the risk of delamination. This includes the use of protective coatings, such as epoxy or silicone which can help to reduce the accumulation of PDs. Additionally, the use of surge arresters, which are designed to limit the severity of transient overvoltages, can help to reduce the risk of delamination.

The thermal cycle can cause delamination in the interface between the conductor and the main insulation. This delamination process can result in PDs that can quickly lead to failure, especially in coils of multiple spirals. Figure 2.8 shows a transverse section of a stator bar, where internal delaminations in the insulation can be observed. These delaminations can cause PDs.

2.4.3 *Slot Discharges*

Slot discharges are PDs which occur in the slot of electric machines. This particular type of discharge is considered to be one of the most damaging for electric machines, as it can cause deterioration of both the insulation material and the copper conductor in the slot, thereby leading to an increase in the temperature of the conductor and consequently accelerating its aging.

Slot discharges in hydrogenerators are a common phenomenon in systems with large air gaps, such as those found in hydroelectric power plants. A slot discharge occurs when the air gap between the rotor and stator is too large and electrical arcing occurs between the two components. This can lead to a significant loss of power and can even damage the generator components.

The size of the air gap between the rotor and stator is determined by the design of the generator and the amount of cooling air that is allowed to flow through the system. If the air gap is too large, the electric field will be strong enough to cause electrical arcing. This arcing will cause the voltage on the rotor and stator to become out of balance, resulting in a slot discharge.

The energy of slot discharges is usually higher than that of other types of PDs. They are often caused by the presence of air gaps, metal particles, water and other objects in the slot of the electric machine's stator. Slot discharges can also be caused by voids in the insulation material, which may be due to aging or the accumulation of dirt and dust.

Slot discharges cause a significant loss of power and can damage the generator components. In extreme cases, the rotor can be melted or even thrown from its bearings. To reduce the risk of slot discharges, the air gap must be kept within the recommended range for the generator. Additionally, protective measures such as cooling fans, shields and insulation can be installed to prevent arcing and reduce the chances of a slot discharge.

Slot discharges in hydrogenerators can occur when the coating of the semi-conductive part in the slot is damaged due to the movement of the bar or the coil in the slot. High levels of discharges will appear when serious mechanical damage is already present, which could result in additional damage to the main insulation and eventually lead to an insulation failure. Generally, slot discharges are caused by the concentration of local electric field which occurs only in the high voltage terminals of each phase.

The absolute time between the detection of PDs and total failure of insulation is often unknown. However, compared to other types of deterioration, this time is typically shorter, particularly in the presence of vibration of bars or coils. Therefore, reliable detection of the early stages is essential to decide which maintenance actions must be taken.

Slot discharges can be prevented by careful design of the generator and by using protective measures, such as insulation and cooling fans. Regular maintenance and inspections should also be performed to ensure the air gap is kept within the recommended range. In order to reduce the occurrence of slot discharges, it is essential to ensure proper maintenance of electric machines, including periodic cleaning of the slot and insulation material, as well as the use of high-quality insulation material.

2.4.4 Discharges in the End-Winding (Corona)

The discharges in the end-winding (corona) of an electrical machine are an important aspect of the operation of the machine. Corona discharges are generated when the electric field between the end-winding and the adjacent surfaces exceeds the dielectric strength of the air surrounding the end-winding. These discharges are associated with the creation of ozone, which can cause damage to the insulation of the machine. To minimize the corona discharges, special care must be taken in the design of the end-winding and the adjacent surfaces. It is usually initiated when an electric field near a conductor reaches a certain threshold value. In hydrogenerators, this happens when the electric field near the rotor is large enough to cause ionization of the air molecules.

It is an electro-hydrodynamic process which involves the flow of ions and electrons between the stator and the air. This is accompanied by an electric breakdown of the air molecules which leads to the formation of an arc between the stator and the air. This arc is responsible for the bright glow and the buzzing noise which is associated with the corona discharge. The formation of stator corona discharge is primarily caused by the electric field that is created between the stator and the rotor of the hydrogenerator. The electric field is caused by the voltage differences between the two components. This field can be increased by increasing the voltage difference, or by introducing stray capacitance in the form of air gaps or insulation defects. Stator corona discharge can also be caused by a combination of some factors. They can include high humidity, high temperatures and the presence of dust or other contaminants in the air. These contaminants can increase the electric field strength and reduce the breakdown voltage of the insulation material allowing the formation of the stator corona discharge.

Corona discharges in the coil head area may occur at various locations with a high electric field concentration. These discharges usually happen at the interfaces between different elements of the stator winding coil head. The main cause of this type of discharge is the presence of a non-uniform electric field in the winding head. This non-uniformity can be attributed to the presence of air gaps, insulation layers, and differences in the turns diameter. Corona discharges are characterized by the presence of high-frequency oscillations, which can be discriminated by the PD pattern signature. This signature is characterized by the presence of narrow pulses with a broad spectrum of frequencies and high amplitudes.

The physics of corona discharges is a complex phenomenon that involves many factors. The electric field between the end-winding and the adjacent surfaces is a key factor in the generation of corona discharges. The electric field is affected by the voltage applied to the end-winding, the size of the end-winding, the shape of the end-winding and the materials used in the construction of the end-winding and the adjacent surfaces.

Conductive particles are one of the main sources corona PDs. These particles can be found in the form of dust, small pieces of metal, or graphite produced by external sources, e.g. wear of mechanical parts of rotating machines. The presence of these particles results in a decrease of the insulating performance of the system and can even cause damages to the insulation itself.

In order to minimize the generation of corona discharges, it is important to design the end-winding and the adjacent surfaces with the appropriate size and shape, to use materials with high dielectric strength and to ensure that the electric field between the end-winding and the adjacent surfaces does not exceed the dielectric strength of the air. Furthermore, it is also important to maintain the air temperature, pressure, and relative humidity at appropriate levels to minimize the likelihood of corona discharges.

2.4.5 *Surface Tracking*

Stator surface discharge, also known as surface tracking, is a phenomenon which occurs when high voltage electricity is conducted through the stator surface of a hydrogenerator. This phenomenon can cause insulation damage, arcing and even damage to the stator winding. It is important to be aware of the risk of stator surface discharges, as well as the necessary precautions to take to prevent them.

Stator surface discharges are primarily caused by the high voltage differential between the stator surface and the surrounding environment. This voltage gradient can cause a spark to jump from the stator surface to the atmosphere, resulting in a discharge. In addition, the presence of dust and other particles can also increase the likelihood of a discharge.

Surface discharges typically start when the electric field along the surface exceeds the gas or liquid dielectric rigidity value. This phenomenon may occur when no coating for stress control is applied in the coil head, or when the coating applied becomes ineffective due to contamination by dirt or humidity, porosity, and thermal effects, among others.

Surface discharges are usually caused by the formation of a track along the surface of the insulation material. This track is formed by a succession of electric sparks of small duration, typically 0.3–3 μs, at regular intervals and following a certain trajectory. Surface discharges can cause accelerated deterioration of the insulation material, leading to insulation failure and, consequently, faults. This type of discharge is normally a very slow failure mechanism, even though the behavior of the PD can change relatively quickly due to surface effects. Surface discharges usually result in a phase-to-ground fault.

To prevent stator surface discharges, it is important to ensure that the stator surface is properly insulated. This can be achieved through the use of insulating materials such as epoxy paint or mica tape. Additionally, it is important to keep the stator surfaces free of dust and other particulates by regularly cleaning and inspecting the surfaces. It is also important to monitor the stator surfaces for signs of damage or wear. This can include checking for cracks, discoloration or other irregularities. If any of these signs are present, they should be addressed as soon as possible to prevent further damage.

PRPDs of stator surface discharges (surface tracking) are a relatively new field of study, but one which is of great importance for the diagnosis and protection of hydrogenerators. This technology enables the detection of PDs on the stator surface, allowing for the tracking of their origin and the determination of their impact on the generator. This technology can also be used to detect any changes in the conditions of the stator windings, allowing for timely diagnosis and maintenance of the system. The use of PRPDs of stator surface discharges is an important step in the diagnosis and protection of hydrogenerators. By using this technology, it is possible to identify any issues in the stator windings, allowing for timely maintenance and repairs. Furthermore, the technology can also be used to protect the generator against any sudden changes in the conditions of the stator, thus ensuring the stability of the system.

In addition to insulation and regular maintenance, it is important to ensure that the hydrogenerator is operated within its rated capacity. If the generator is operating above its rated capacity, this can lead to an increase in the voltage differential between the stator surface and the atmosphere, resulting in an increased risk of discharges.

Stator surface discharges can lead to the formation of sulfuric acid which can corrode the stator windings and other components of the generator. This acid can also migrate through the insulation of the stator leading to further damage and decreased reliability. Additionally, the formation of sulfuric acid can lead to the formation of hydrogen sulfide, which can accumulate in the generator's stator casing and cause damage to the insulation and other components.

Furthermore, the formation of sulfuric acid can lead to the breakdown of organic materials, such as the insulation material of the stator windings. This breakdown can lead to the formation of combustible gases, such as carbon monoxide and hydrogen, which can lead to a risk of fire or explosion. Additionally, the breakdown of organic materials can lead to the formation of highly corrosive gases, such as formaldehyde, which can cause further damage to the generator's components.

Stator surface discharges can also lead to the formation of other chemical compounds, such as calcium sulfate, which can accumulate on the stator windings and cause further damage. Additionally, the accumulation of these compounds can lead to the formation of an insulating layer, which can lead to decreased electrical performance and increased risk of failure. Therefore, it is important for operators to regularly inspect the generator for signs of stator surface discharges and take steps to mitigate the risk of chemical damage.

Finally, it is important to ensure that all personnel working on or near the hydrogenerator are aware of the risk of stator surface discharges. Personnel should be instructed to follow all safety protocols and to take extra precautions when working in close proximity to the stator surface.

2.4.6 Phase to Phase Discharges (Gap Discharges)

Phase to phase discharges, or gap discharges, are a type of electric discharge that occurs when enough energy is stored in the air gap between two or more phase conductors. This energy is caused by the buildup of an electric field due to the difference in potential between the two phases. When the electric field is strong enough, a spark will occur resulting in an electric discharge. This type of discharge is most common in hydrogenerators, where the high voltage created by the generator is transferred between the phases.

Thus, phase to phase discharges occur in the gaps between two conductors in the same electric phase. This type of discharge is characterized by the presence of lower magnitudes of electric current and higher voltages, making it more dangerous than conductor discharges.

In terms of physical and electrical parameters, the gap discharges are characterized by an amplitude of current that usually oscillates between few hundreds of picoamperes to hundreds of nanoamperes. Furthermore, the voltage involved typically ranges from few tens of kilovolts to hundreds of kilovolts.

PDs can occur between phases due to inadequate insulation distance or inadequate usage of support systems of the coil head. Depending on the project, these discharges can have high magnitudes and can manifest as surface discharges or internal discharges. In some cases, this can lead to a phase-to-phase fault.

The phase discharges can be divided into two categories: thermal and mechanical effects. The thermal effects result from the intense heat created by the spark, which can cause physical damage to the surrounding materials. This can range from melting insulation to burning the surrounding metal. The mechanical effects result from the shock wave created by the spark, which can cause physical damage such as cracking or breaking of the insulation, or even shaking of the surrounding structure. If the spark is strong enough, it can cause fire, which can quickly spread to other parts of the hydrogenerator. The shock wave created by the spark can also cause more serious damage, such as damaging the turbine blades or even damaging the generator itself.

It is important to take measures to reduce the buildup of electric fields. This can be done by installing surge suppression devices which detect the buildup of electric fields and discharge energy before a spark occurs. Other measures, such as increasing the gap between the phases or increasing the insulation, can also help to reduce the risk of spark occurrences.

The radiation from phase to phase discharges is typically in the region of radio frequencies. This type of radiation can be detected with specialized equipment. The amount of radiation depends on the voltage between the two phases and the length of the gap between them. High voltage systems with longer gaps will generate more radiation than low voltage systems with shorter gaps.

The amount of radiation generated by phase to phase discharges can be further increased if the spark is allowed to persist for a long period of time. In this case, the spark can reach temperatures of several thousand degrees Celsius and emit significant amounts of infrared radiation. This radiation can be even more hazardous than the radio frequency radiation and should be avoided.

In addition to the measures mentioned above, it is also important to regularly inspect and maintain the hydrogenerator to ensure that the electrical components are in good condition. This will help to reduce the risk of a spark occurring. Regularly inspecting the hydrogenerator will also help to identify any potential problems before they become serious, and allow any necessary repairs to be carried out in a timely manner. Hydrogenerators are particularly susceptible to phase to phase discharges due to their large size and the high voltages that they must handle. Special attention must be taken to ensure that all of the components of a hydrogenerator are properly insulated from each other to reduce the risk of a phase to phase discharge occurring.

References

1. Sigmond, R. S. (1984). The residual streamer channel: Return strokes and secondary streamers. *Journal of Applied Physics, 56*(5), 1355–1370.
2. Naidu, M. S., & Kamaraju, V. (1996). *High voltage engineering* (2nd ed.). New York: McGraw-Hill.
3. Stone, G., Culbert, I., Boulter, E., & Dhirani, H. (2014). *Electrical insulation for rotating machines: Design, evaluation, aging, testing, and repair* (2nd ed.). Piscataway: Wiley-IEEE Press.
4. Institute of Electrical and Electronics Engineers. (Dec. 2014). IEEE Guide for the Measurement of Partial Discharges in AC Electric Machinery. IEEE Std 1434-2014 (Revision of IEEE Std 1434-2000) (pp. 1–89).
5. Stone, G., & Kapler, J. (1998). Stator winding monitoring. *IEEE Industry Applications Magazine, 4*(5), 15–20.
6. Bartnikas, R., & McMahon, E. (1979). *Engineering dielectrics: Corona measurement and interpretation - STP 669* (Vol. 1). ASTM special technical publication. West Conshohocken: American Society for Testing and Materials.
7. International Electrotechnical Commission. (2000). High-voltage test techniques-partial discharge measurements. IEC, Publication-60270.
8. Rotating Electrical Machines - Part 27-2: On-Line Partial Discharge Measurements on the Stator Winding Insulation of Rotating Electrical Machines. IEC/TS 60034-27-2. International Electrotechnical Commission. Tech. Rep. (2012).
9. Lyles, J., Stone, G., & Kurtz, M. (1988). Experience with PDA diagnostic testing on hydraulic generators. *IEEE Transactions on Energy Conversion, 3*(4), 824–832.
10. Stone, G. C., Boulter, E. A., Culbert, I., & Dhirani, H. (2003). Rotating machine insulation systems. In *Electrical insulation for rotating machines: Design, evaluation, aging, testing, and repair* (pp. 1–41). John Wiley & Sons, Ltd. https://doi.org/10.1002/047168290X
11. Eichwald, O., Ducasse, O., Dubois, D., Abahazem, A., Merbahi, N., Benhenni, M., & Yousfi, M. (Nov. 2008). Experimental analysis and modelling of positive streamer in air: Towards an estimation of O and N radical production. *Journal of Physics D: Applied Physics, 41*(23), 234 002.

Chapter 3
Partial Discharges: Frequency Characteristics, Sensors and Laboratory Measurements

3.1 Types of PD Sensors

PDs can cause serious damage to electrical components and systems and the detection of PDs is essential for the reliable operation of high voltage electrical installations. Several types of sensors can be used to detect PDs. These sensors can be classified according to the type of signal they measure, and the most common types are acoustic, optical, ozone and electric sensors. Each type of sensor has its own advantages and disadvantages, and the appropriate type should be selected for each application.

Acoustic Sensors
Acoustic sensors measure sound waves, usually ultrasound, generated by PD events. They can be used to detect PDs in a wide range of high voltage equipment and environmental conditions. The advantages of acoustic sensors include their low cost, low sensitivity to electromagnetic interference and their ability to detect PDs in areas that are difficult to access using other types of sensors. On the other hand, they are not very precise in locating the source of the discharges.

Optical Sensors
Optical sensors measure light emitted by PDs. They are often used to detect PDs in high-voltage equipment, such as generators and cables. The advantages of optical sensors include their high sensitivity and accuracy in locating the source of the discharge. They are also relatively unaffected by electromagnetic interference. On the other hand, they are more expensive than other types of sensors and may not be suitable for some applications.

Ozone Sensors
Ozone sensors can be used for detecting partial discharges in hydrogenerators. These sensors have the ability to detect ozone concentrations in the air and can be used to detect the presence of partial discharges that can occur in hydrogenerators.

V. Dmitriev et al., *Partial Discharges in Hydroelectric Generators*, Power Systems,
https://doi.org/10.1007/978-3-031-36604-8_3

Partial discharges ionize air and this process involves chemical reactions from which ozone is produced. With an ozone sensor, the presence of partial discharges can be easily detected.

The ozone sensor works by measuring the ozone concentration in the air near the generator. Ultraviolet sensors are able to detect even small amounts of ozone, so it is able to detect very small discharges. Ozone concentrations in the air are also typically higher near the generator. On the other hand, the more affordable electrochemical sensors can be used, but typically only ozone concentrations over 100 ppb can be detected (much higher concentration detection threshold than those presented by ultraviolet instruments).

Ozone sensors are a reliable and cost-effective solution for detecting partial discharges in hydrogenerators. The sensors are relatively easy to install and can be used to detect discharges that could cause significant damage to hydrogenerators. Inspection and maintenance procedures are required to keep ozone sensors working because their responses are variable over time since they are sensitive to environmental conditions, such as humidity and temperature.

Electric Sensors

Electric sensors measure electric fields generated by PDs. They are normally used to monitor high-voltage equipment, such as generators and cables and can detect PDs in a variety of media, including air, oil, and solid insulation. The advantages of electric sensors include their high sensitivity and accuracy in locating the source of the discharge. However, they are more expensive than acoustic sensors and are more susceptible to electromagnetic interference.

At first, PDs can be detected using either pulsed-driven or electromagnetically irradiated signals. Sensors that detect pulsed-driven signals typically consist of high voltage capacitors coupled in series with low voltage devices; this arrangement is known in the literature as capacitive coupling. Sensors that detect electromagnetically irradiated pulse signals are typically antennas that exhibit significant sensitivity depending on their installation location and frequency response function.

Directional couplers are also used for PD detection in hydrogenerators. These devices can be installed in the generator and measure the PD signals propagating in a particular direction. It is a proximity electromagnetic detection device, so it should be installed on a surface of interest. Multiple directional couplers can be used to measure PD signals propagating from different directions at the same time, which can be useful if multiple sources of PDs are present. Directional couplers are wideband devices and, thus, can provide detailed information about PD pulses.

3.2 Spectral Characteristics of Electromagnetic PD Signals at Source and at Stator Measuring Terminals

Figure 3.1 schematically shows frequency responses of an ideal PD pulse at the discharge source (superior cut-off frequency f_{uDPo}) and its ideal frequency

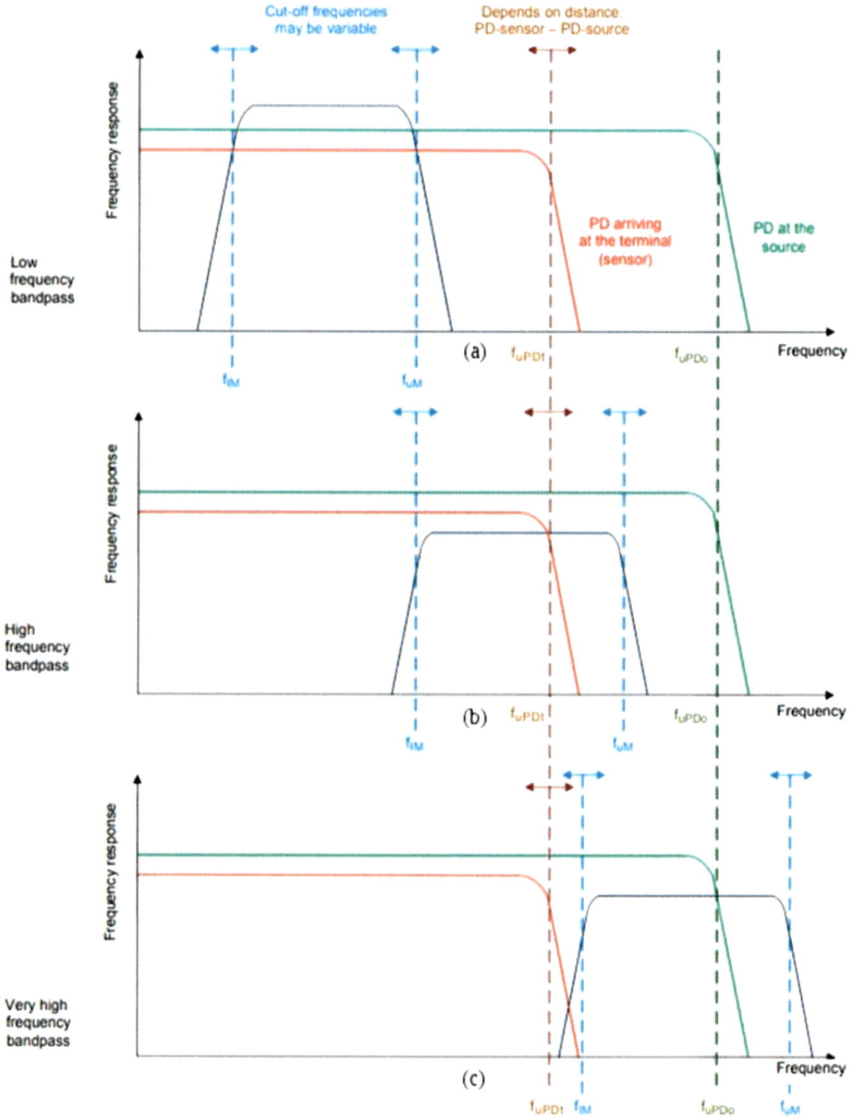

Fig. 3.1 Typical frequency responses of PD pulses obtained at the PD source and at machine terminals for different measuring systems: (**a**) low frequencies range, (**b**) high frequencies range and (**c**) very high frequencies range. Source: adapted from [1]

response at the machine terminals (superior cut-off frequency f_{uDPt}). Notice that $f_{uDPt} < f_{uDPo}$ because the winding path between PD excitation region and machine terminals acts as a propagation channel filtering high-frequency energy of the electromagnetic waves.

PD measuring systems consisting of PD sensors, cables and measurement instruments, usually include band-pass filters with specific superior and inferior cut-off frequencies, which depend on the sensor design and on internal impedance of measuring instruments. In Fig. 3.1a, b and c, we have three examples didactically showing the frequency responses of different measuring instruments. The cut-off frequencies and bandwidths of commercially available systems can vary greatly. The characteristics of the frequency response of a complete measurement system have a considerable impact on the global detection sensibility and on the acquired signal properties. It should be noticed that Fig. 3.1 solely presents the fundamental characteritics of ideal curves. Depending on the winding design and on the used measuring system in hydrogenerator field cases, we may have diverse effects affecting the exact shape of the frequency response curves, thus affecting the measured PD signals.

Low-Frequency Range

The low-frequency range is used to measure PDs in insulated bars, coils and components of the generator [1]. The measurements of PDs performed in this range of frequencies offer the possibility of detecting PDs located far away from the PD sensor. However, this range is subjected to noise and disturbances present during online measurements, requiring special procedures for performing separation of noise and disturbances from PD signals. Measurements conducted with low cut-off frequencies, normally between 100 kHz and 3 MHz, allow for good sensitivity not only to PDs near the PD sensor but also to PDs located further away in the winding.

The measurement of conduction electrical signals produced by partial discharges occurring in hydrogenerator windings in the frequency range between 100 kHz and 3 MHz is a critical task in order to prevent potential damage to the winding. This is achieved mainly by coupling capacitors to the winding in order to measure the voltage levels of the signals. The capacitors act as a medium for the transmission of PD electrical energy from the winding to the sensors, providing accurate measures of the electrical signals generated by the PDs. In order to accurately measure the PD conduction signals, the coupling capacitors and other sensors must be carefully designed and tested. The capacitance value of the coupling capacitor must be selected according to the frequency range of the PD signals to be measured, in order to ensure that the transmission of energy is minimally distorted.

Measuring instruments that work in low-frequency ranges are primarily used to detect information from the constant part of the spectrum of PD pulses (see Fig. 3.1a). Considering that the upper cutoff frequency of the measuring instrument's passband is significantly lower than the upper cutoff frequency of the frequency response of the PD pulse, the detected pulses are directly proportional to the currents of the PD pulses.

Once the coupling capacitors and other devices have been properly installed, they can be used to measure the conduction electric signals produced by the partial discharges and to monitor the condition of the winding. This monitoring can then be used to identify any potential problems with the winding and to take preventive measures to ensure the safety of the hydrogenerator.

High Frequency Range
This range is used for the study of PDs above 1 MHz, and it is the best range for the characterization of PDs with repetition rate greater than 1 kHz, because of the observed good sensibilities to the PDs high peak voltages. In this range, the detection criterion is usually based on the peak voltage, which must be greater than a pre-determined level, depending on the specific system used. The high frequency range is considered to be between 3 MHz and 30 MHz. The lower cut-off frequencies can also be tuned below 1 MHz to ensure good sensitivity on the measurement. However, generally, lower cut-off frequencies above 1 MHz are used to improve the suppression of noise signals present in the low frequency range.

The detection of PDs at high frequencies is less susceptible to noise and can be used to characterize the PD faults based on pulses that reach the sensor based on their pulse shape. Therefore, this allows for discrimination between different PD sources, according to the signal shape. When the upper cutoff frequency of the measuring system is much higher than the upper cutoff frequency of the PD pulse that reaches the sensor, the pulse will no longer be directly proportional to the apparent load of the PD pulse. Results of PD measurements in the high frequency range are usually expressed in voltage units, such as volts V or millivolts mV.

Very High Frequencies Range
In this range, one employs a typical bandwidth of several hundred MHz: lower and upper cutoff frequencies are typically 30 and 300 MHz, respectively. The phenomenon of PDs can be detected in this range by measuring the frequency spectrum. In this range, the electrical equipment noise can be higher than the PD radiated signal, resulting in a lower signal-to-noise ratio (SNR) if adequate procedures are not taken.

It can be seen from Fig. 3.1, the frequency response of measuring system exhibits pronounced overlaps with the frequency response of the original PD pulse. Consequently, measurements conducted in high frequency ranges can have good sensitivity to signals generated near the PD sensor.

The PD sensor must be installed in the high voltage terminals, ideally close to the bars with higher electrical stress in the winding, in order to improve signal/noise ratio and to reduce susceptibility to external noise. As the upper cutoff frequency of the measuring system is higher than the upper cutoff frequency of the PD pulse arriving at the sensor, the pulse is no longer directly proportional to the apparent PD pulse load. The results of PD measurements in the high-frequency range are usually expressed in voltage units.

Ultra High Frequencies Range
In this range, inferior cut-off frequencies of approximately 300 MHz and superior cut-off frequencies up to 3 GHz are normally adopted. The PD sensors that work in this frequency range are antennas that detect electromagnetically radiated signals from the PD pulses. The signal energy detected by these sensors as well as the measuring sensitivity mainly depend on the antenna location, the distance between the antenna and the PD source and the measurement system bandwidth.

In the UHF range, the PD signals are usually relatively strong, but the UHF sensors are more sensitive to the electromagnetic interference than the PD sensors in the lower frequency ranges. Notice that several telecommunications services operate in the UHF region. Furthermore, the UHF range presents a higher attenuation of the signals. Characteristics of the measuring system, specially antenna impedance, radiation pattern and efficiency have to be carefully designed to improve the PD signal detection. Microstrip couplers may also be used in this range.

3.3 Electromagnetic Couplers

3.3.1 Stator Slot Coupler (SSC)

The stator slot coupler (SSC) is a commonly used sensor for detecting PDs in rotating machines. It consists of a small low-voltage sensor installed in the stator slot of the machine. The SSCs are designed to be used in high-voltage machines operating up to 40 kV.

SSCs are essentially directional couplers, which work in the frequency range from 30 MHz to 1 GHz, according to the standard [2]. The standard [3] emphasizes that the SSC should be installed as close as possible to the winding region that is most prone to PD activity. The SSC shown in Fig. 3.2 consists of a ground plane and a sensor line with coaxial cables at each end of the output. Whenever an electromagnetic wave, such as a wave excited by a PD pulse, propagates along the SSC near the sensor line, an output pulse is detected. Thus, SSCs can be used to identify and measure PD signals. This method is based on the measurement of the voltage signal between the ground plane and the line. The signal is measured by using an oscilloscope or a spectral analyzer, for example. Through this method, the PD signals can be detected, measured and used to analyze the insulation systems of HV machines.

The characteristic impedance of SSCs is typically 50 Ω, which matches the impedance of the standard output coaxial cables used in signal meters. The presence of two outputs allows one to determine the direction of propagation of the PD pulse and, potentially, its location. This is possible by using instrumentation capable of

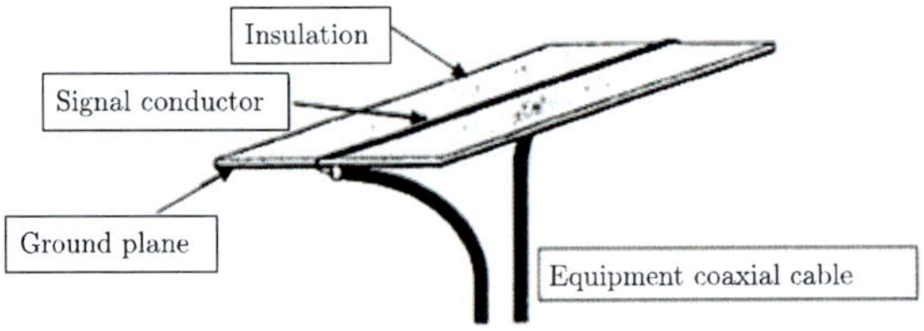

Fig. 3.2 Simple SSC scheme. Source: adapted from [4]

measuring at which end of the SSC the first signal is detected. One of the main advantages of a SSC is its wide operation band.

3.3.2 Practical Realization of SSC-Microstrip Directional Coupler

A microstrip directional coupler is a device used to couple electric energy between two microstrip lines. It is used in high-frequency applications such as wireless communication networks and radar systems. A common microstrip directional coupler consists of two parallel microstrip lines and a coupling region that connects them. The coupling region is designed to induce a voltage in the second line proportional to the voltage in the first line. Thus, this voltage can be used to measure the magnitude and phase of the electric signal in the first line. The first line in the SSC coupler is the stator bar itself and the second line is the microstrip line placed close to the bar. The microstrip directional coupler can be used in high frequencies up to tens of GHz.

A microstrip directional coupler was developed using a duroid substrate. The coupler surface area measures $250\times20\,\text{mm}^2$ with a substrate thickness of 1.5 mm. The width of the feed line is 3 mm to provide a $50\,\Omega$ impedance. The measured results of the developed directional coupler show that it has excellent return loss making it suitable for sensor applications. Figure 3.3 shows a fabricated prototype of the microstrip line coupler used in this work experiments to measure PD signals. As shown in Fig. 3.4, the experimental results of the return loss are in accordance with the simulations with obtained levels below -10 dB over the entire measured frequency range (90–20,000) MHz. The coupler was successfully used to measure PD in high voltage equipment.

3.3.3 Loop Antennas

Loop antennas have been used for the detection of PDs, according to works published by Rozi and Khayam [5, 6]. In our project, a loop antenna has been developed for measurements of PDs. Figure 3.5a shows the antenna's dimensions, and Fig. 3.5b displays the fabricated prototype. The return loss measurement, depicted in Fig. 3.6, indicates that the loop antenna has good response for frequencies up to approximately 850 MHz (and reasonable response from 850 MHz to 2 GHz). The radiation pattern for 300 MHz is shown in Fig. 3.7, which reveals the antenna's directionality.

The loop antenna is connected to a low noise amplifier whose gain is adjustable, and its output is connected to a spectrum analyzer, which is used to analyze the signal, or to an oscilloscope. The gain of the amplifier is adjusted in order to provide

Fig. 3.3 Fabricated prototype of microstrip directional coupler: (**a**) microstrip directional coupler and (**b**) encapsulated microstrip directional coupler

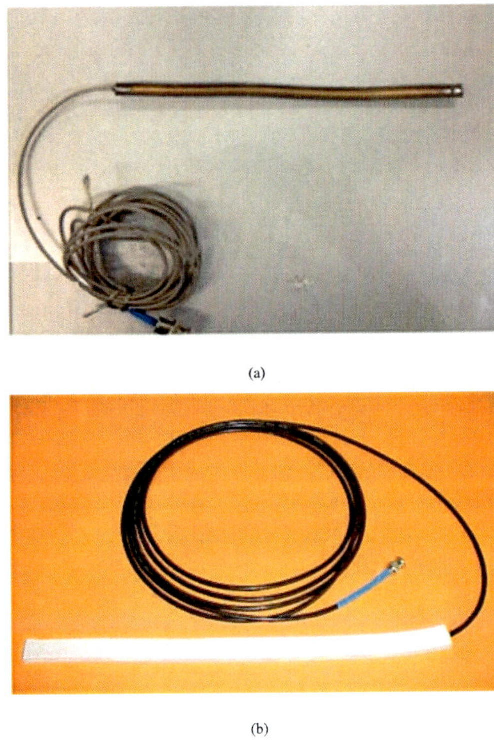

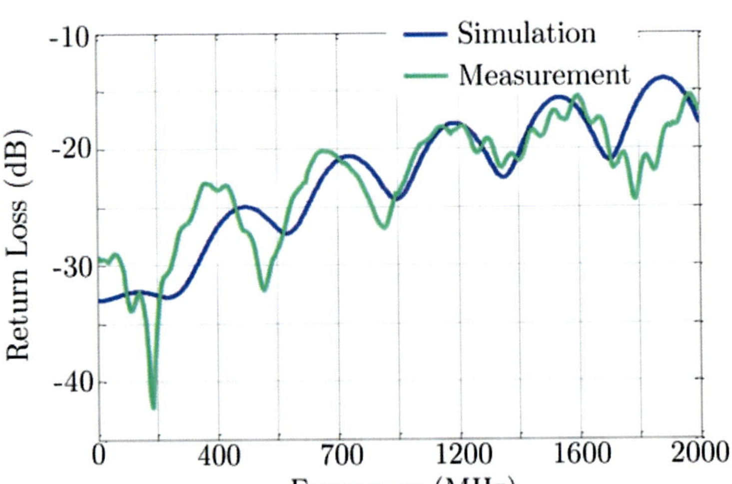

Fig. 3.4 Simulated and measured return losses of microstrip directional coupler

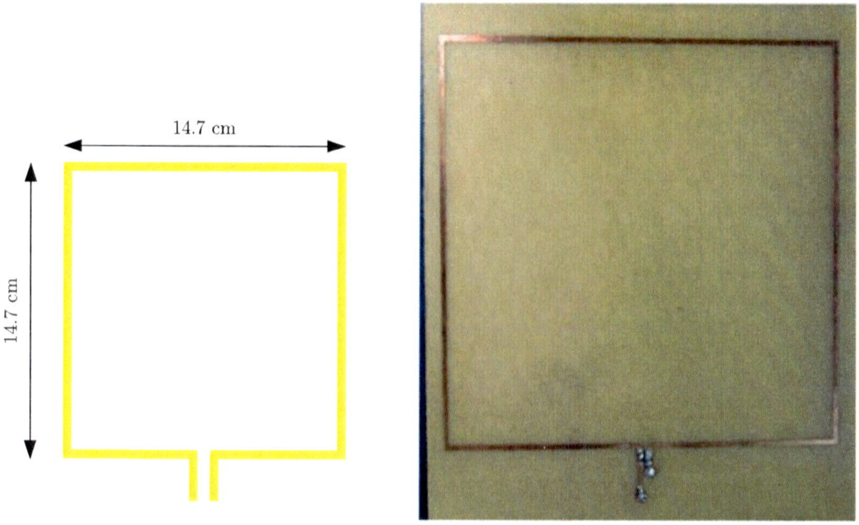

(a) (b)

Fig. 3.5 Loop antenna: (**a**) geometry and (**b**) the manufactured prototype

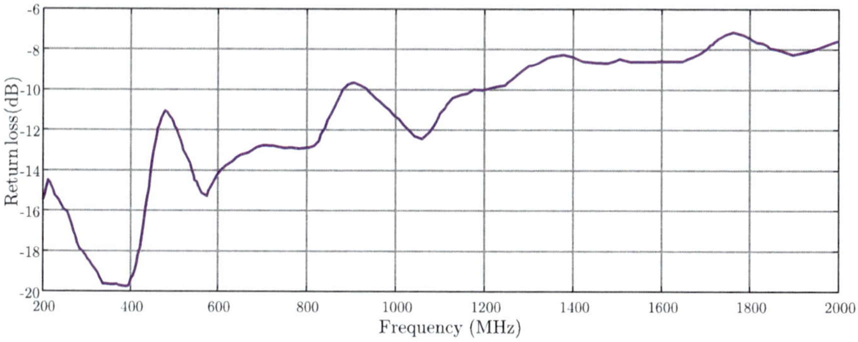

Fig. 3.6 Measured return loss of prototype loop antenna

an adequate signal-to-noise ratio, allowing for the detection of PDs with low power levels. The spectrum analyzer can be used to detect the presence of PDs, as well as to measure power. Time domain characteristics, such as PD pulse shape and PD peak phase placement (as in phase-resolved partial-discharge map, PRPD), are usually analyzed using a digital oscilloscope and specially developed software. Finally, time-frequency PD maps have also been done using spectrum analyzer [7].

Figure 3.6 presents the return loss measured for the loop antenna. This antenna has a frequency range of 0.2 GHz–2 GHz, which covers a wide part of the PD frequency spectrum. The antenna has a return loss of −7.0 dB or less, which makes it suitable for PD detection. Figure 3.7 shows the radiation patterns of the

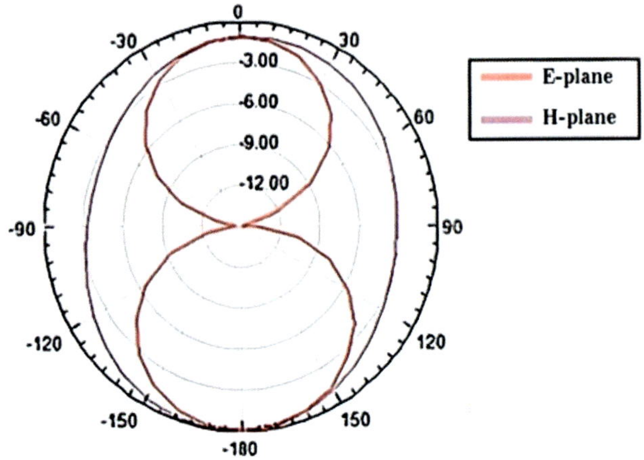

Fig. 3.7 Radiation patterns of loop antenna (two-dimensional polar plots of E-plane and H-plane) measured at 300 MHz

loop antenna, represented in a two-dimensional polar plot obtained at 300 MHz. The loop antenna is an interesting candidate device for detecting PDs, since it is relatively inexpensive and easy to construct [6, 8]. Its E-plane radiation pattern [9] is composed of two opposite lobes. This radiation pattern is useful for detecting PDs in different directions, allowing the user to direct the antenna towards the PD source, possibly reducing effects of external interference sources in certain situations.

3.3.4 *Log-Periodic Antenna*

A log-periodic antenna (LPA) is an antenna that exhibits a log-periodic frequency response, which means that the impedance and radiation patterns of the antenna remain approximately constant over a wide frequency range. This type of antenna is particularly useful for measuring PDs in high-voltage equipment. LPA is also used for PD detection in other applications such as in aircrafts and automobiles.

The LPA consists of a series of dipole elements connected to feedline with phases in alternance. Each dipole has a different length and spacing to neighbour dipoles. The spacing between the elements is logarithmically proportional to the frequency, and the length of the dipole element is inversely proportional to the frequency. This arrangement creates a wide bandwidth response with a practically constant impedance and radiation pattern over a wide frequency range. The antenna typically has a bandwidth of 10:1 or more which is interesting for PD detection.

The LPA is a very effective tool for PD detection. The antenna can detect PD signals in the range of 10 MHz to 1300 MHz (or up to 2000 MHz if impedance matching requirements can be relaxed), allowing for the detection of PDs in a wide range of applications. In addition, the LPA can be used in both the time and frequency domain measurements (as any other antenna), allowing the user to detect the presence of PDs using temporal or spectral instrumentation. The antenna also has a low noise floor and a high signal-to-noise ratio, allowing for the reliable detection of PDs.

In order to make comparisons between the measurements made by the directional coupler and the loop antenna, a commercial log-periodic antenna was used, which is shown in Fig. 3.8. The characteristics of the antenna are shown in Table 3.1, as reported in [10] and [11].

Figure 3.9 shows the measured return loss for the log-periodic antenna. The antenna was tested in the frequency range from 200 MHz to 1 GHz, with a step of 20 MHz. The return loss was measured by a vector network analyzer (VNA). It can be seen that the return loss is below −5 dB in the entire frequency range which indicates good performance of the antenna. This result is in agreement with

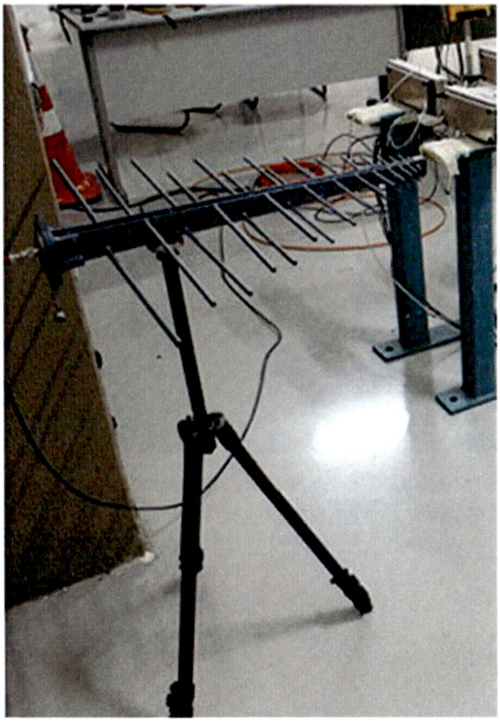

Fig. 3.8 Log-periodic antenna

Table 3.1 Technical specifications of the log-periodic antenna. Source: adapted from [12]

Parameter	Specification
Frequency range	200–1300 MHz
Polarization	Linear
Input impedance	50 Ω
Dimensions	765 mm × 120 mm × 710 mm

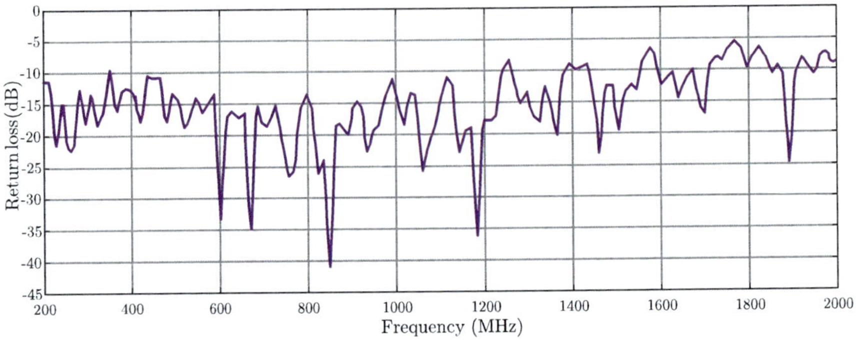

Fig. 3.9 Measured return loss for log-periodic antenna

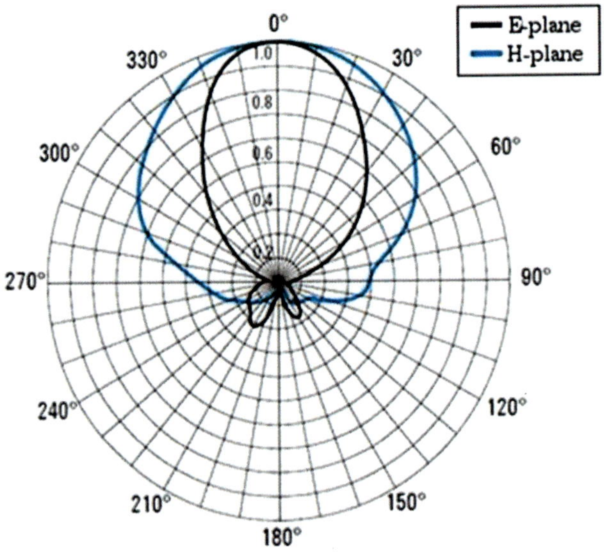

Fig. 3.10 Radiation patterns of the log-periodic antenna measured at 750 MHz (E-plane and H-plane). Source: adapted from [12]

the literature [9]. Therefore, the performance of the antenna is satisfactory for PD applications, which require high-sensitivity measurements.

Figure 3.10 shows the radiation pattern of the log-periodic antenna measured at 750 MHz, adapted from [12]. This type of antenna has a very wide frequency range and its radiation pattern is barely affected by frequency variation, making it suitable for measuring PDs. The antenna, with approximately constant characteristics, can be used for pulse detection in the range of 200 MHz to 2 GHz, with a directive radiation pattern, allowing the detection of short-duration pulses. The antenna is also suitable for detecting PDs at low frequencies, as it has a good signal-to-noise ratio. Additionally, the antenna has a low distortion factor. Directivity is a characteristic which can allow the reduction of external interference if the antenna is properly directed do PDs. The log-periodic antenna is thus an excellent choice for measuring PDs and it is widely used in research and industry. It can provide a reliable performance for pulse detection and can be used to measure the intensity and number of discharges (i.e. PD counting). This makes it an ideal tool for diagnosing and monitoring the health of electrical equipment.

3.4 Experimental Results: Measurements of PD Signals

Different types of discharges found in hydrogenerators (internal discharges, delaminations, corona, slot discharges, surface discharges and gap discharges) were individually reproduced in the laboratory and evaluated using electromagnetic sensors. Signals were compared with the data recorded using the conventional measuring system, which is based on a coupling capacitor. The results obtained in the laboratory indicated that it is possible to accurately characterize different types of discharges according to the UHF frequency components of the signals captured using the electromagnetic sensors. The directional coupler presented the greatest sensitivity in comparison with the loop and log-periodic antennas.

3.4.1 High-Voltage PD Experiments

To analyze PDs in the stator windings, experiments were conducted in a laboratory with an imbricated coil, which is 1.3 m long, taken from a 10 MW hydrogenerator. Figure 3.11 shows the high-voltage experimental setup in Eletronorte HV-Laboratory, Brazil. In the experiments, a 20 kV (60 Hz) source from Hipotronix was used.

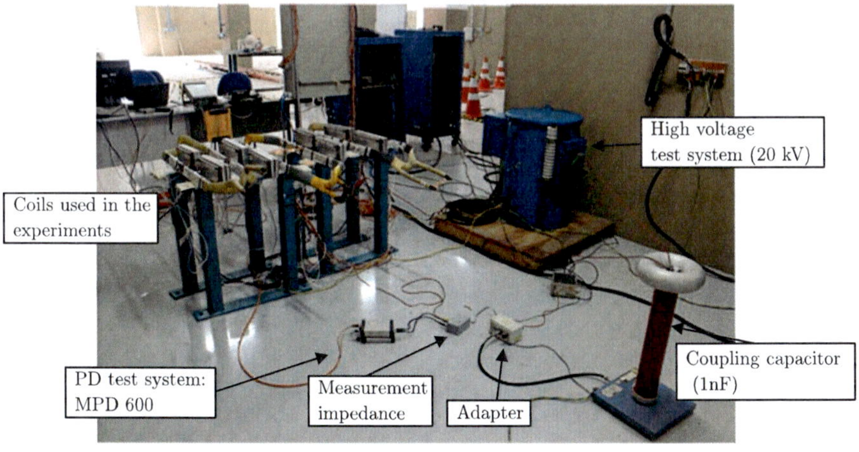

Fig. 3.11 High-voltage laboratory setup

For data collection, two measuring systems were employed: the Monitoring and PD Measuring Device (MPD 600) produced by Omicron and a Tektronix digital oscilloscope MDO 3104 (1 GHz, 5 GS/s). The former was used during experiments for recording PRPD maps. It consists of software and hardware (the hardware is a specific digital oscilloscope). The latter instrument was used to perform time domain records of PD signals.

Figure 3.11 illustrates the experimental setup of the high-voltage laboratory at Eletronorte. The main components of the setup are the hydrogenerator coils, power generator, the high-voltage generator, the capacitors and the discharge electrode. The electric power generator, which can provide up to 8 kV, is connected to the high-voltage generator, which is responsible for increasing the voltage applied to the system up to 40 kV. Measuring system can also be seen, consisting on the MPD 600 test system, a measurement impedance, a connector adapter and an 1 nF coupling capacitor.

The characterization of PD sensors for each source of PD activity was done by comparing the PD pulse waveforms obtained with the different sensors. The following PD sources were analyzed: internal voids, delamination, corona, slot discharges, gap discharges and surface discharges (tracking).

The electromagnetic sensors (directional coupler and antennas) were selected to measure the fields associated to PD activity from the different sources. The PD pulse waveforms obtained using the different sensors were compared with each other. The results obtained from the comparison shows that the directional coupler may be suitable for registering internal PD activity in the stator winding insulation and antennas seems to be applicable for measuring external PD activity (such as corona PDs). The results of this analysis may help in the proper selection of the sensors for the characterization of different PD sources and to monitor PD events in high-voltage systems.

3.4.1.1 Internal Voids

Internal partial discharges are generated in air or gas-filled pockets embedded in the main insulation of power generators. These arise from the manufacturing process, and typically do not cause significant degradation under normal circumstances. Figure 3.12a, b, and c show time domain signals and Fig. 3.13a, b, and c show frequency spectra recorded for internal discharges using a directional coupler, log-periodic antenna, and loop antenna, respectively. It is observed that signals with frequency spectra up to 600 MHz were detected by all three sensors.

Internal voids can be identified in the insulation of the power generators bars using a HV spark test. This laboratory test consists on applying high voltage between bar's central conductor and a grounded metallic part placed over the bar. If the insulation has voids, the sparks will emerge in the voids.

In certain circumstances, internal voids may have a major impact on the power generator's performance. In particular, if the voids have a small size, the electric field near the void is very high and it can cause PDs. Moreover, the electric field

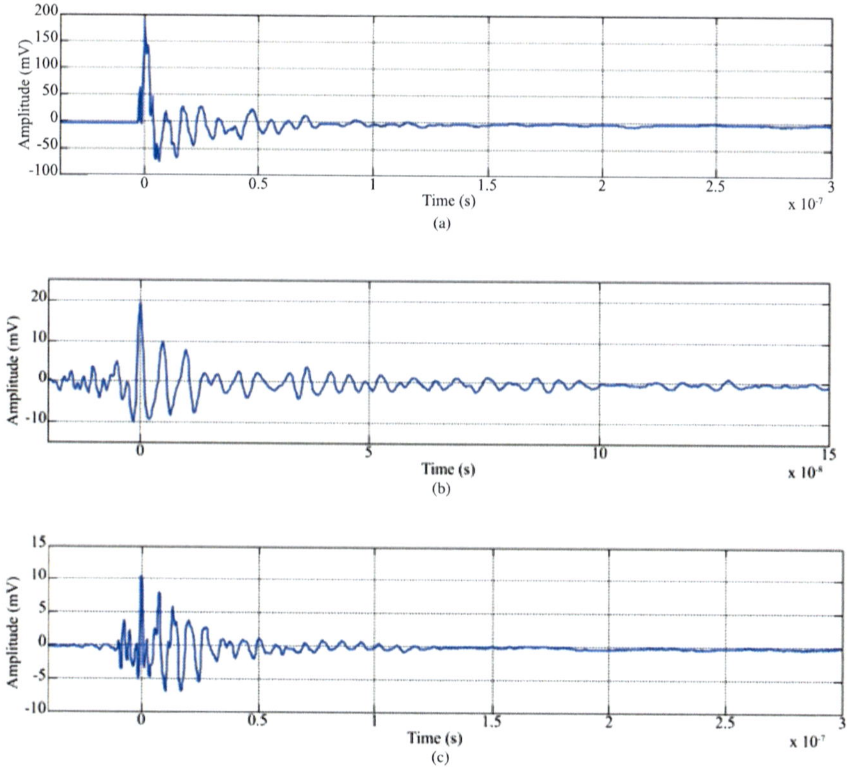

Fig. 3.12 Time domain signals of internal discharges measured using (**a**) directional coupler, (**b**) log-periodic antenna and (**c**) loop antenna

near the void can lead to a fast breakdown of the insulation, resulting in short circuit [3] in severe cases.

The main PRPD characteristic of the resulting partial-discharge pattern for internal discharges is the symmetry between the positive and negative PDs, resulting in rounded shape PD clouds in PRPD map, as illustrated in Fig. 3.14. We can see in that Figure a PD pattern recorded in laboratory when exclusively internal PD activity was present in the bar.

3.4.1.2 Internal Delamination

Internal delamination PDs are generated in air or gas-filled, longitudinally elongated pockets embedded in the main insulation. A typical PRPD pattern resulting from internal delamination is illustrated in Fig. 3.15. After dissection of the coil and visual inspection, the presence of delamination between the insulation layers was verified, as shown in Fig. 3.16. Figure 3.17a, b, and c illustrate signals in the time domain and

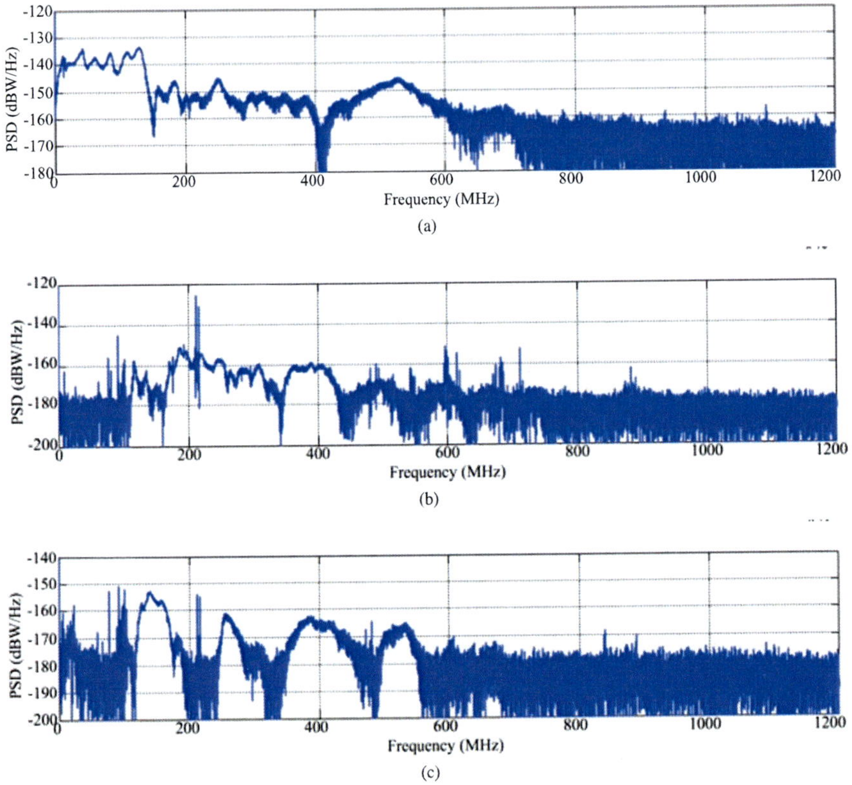

Fig. 3.13 Frequency spectra of internal discharges measured using: (**a**) directional coupler, (**b**) log-periodic antenna and (**c**) loop antenna

Fig. 3.18a, b, and c illustrate frequency spectrum measured by a directional coupler, log-periodic antenna and loop antenna, respectively. It is observed that the three sensors captured signals with frequency spectra up to 800 MHz.

The delamination PD is usually characterized by a triangular PRPD pattern. The occurrence of this PD is usually associated with the presence of cavities or delamination in the insulation (see Fig. 3.16). This defect usually occurs due to mechanical problems such as vibrations, shocks or temperature changes.

Delamination PDs are usually categorized as low-energy PDs, as their magnitude is typically lower than the PDs caused by other types of insulation defects. However, the occurrence of delamination PDs should not be ignored, as they can cause serious damage to the insulation and lead to the failure of the equipment over time [13]. Therefore, the detection and monitoring of delamination PDs is essential for the proper functioning of electrical equipment.

Figure 3.17a, b and c illustrate signals in the time domain and Fig. 3.18a, b, and c illustrate frequency spectra of signals measured using three different

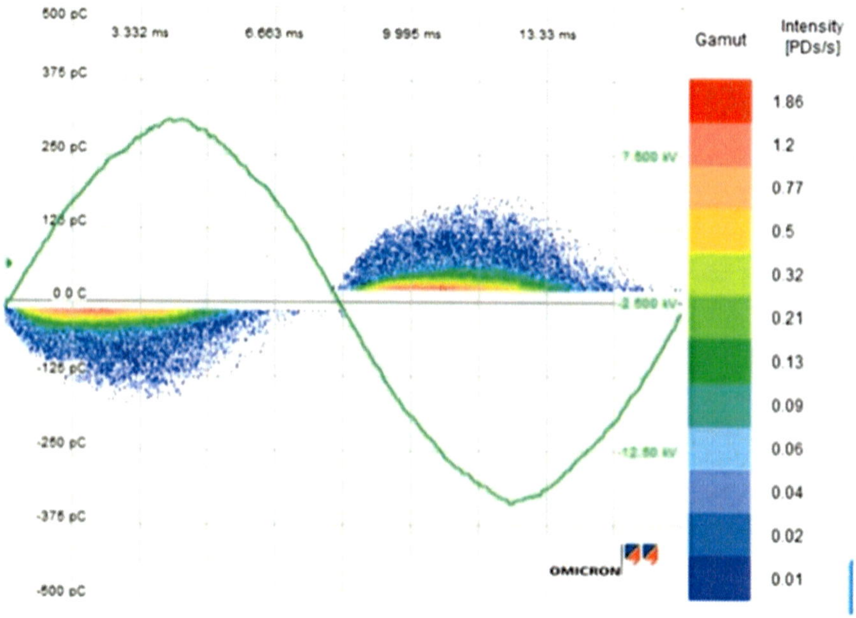

Fig. 3.14 Internal voids: typical PRPD map obtained in laboratory measurement

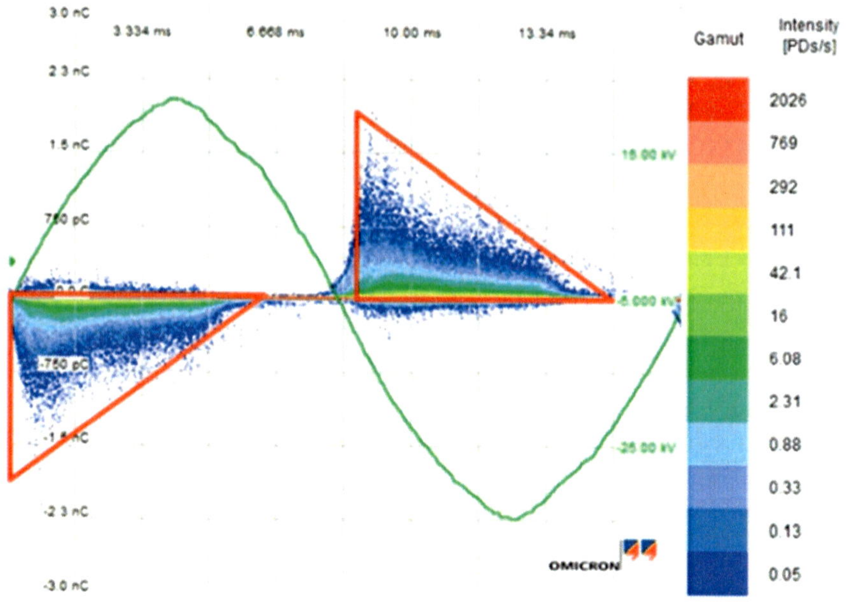

Fig. 3.15 Typical PRPD map obtained in laboratory for delamination

Fig. 3.16 Cross-section of the coil and physical evidences of delamination

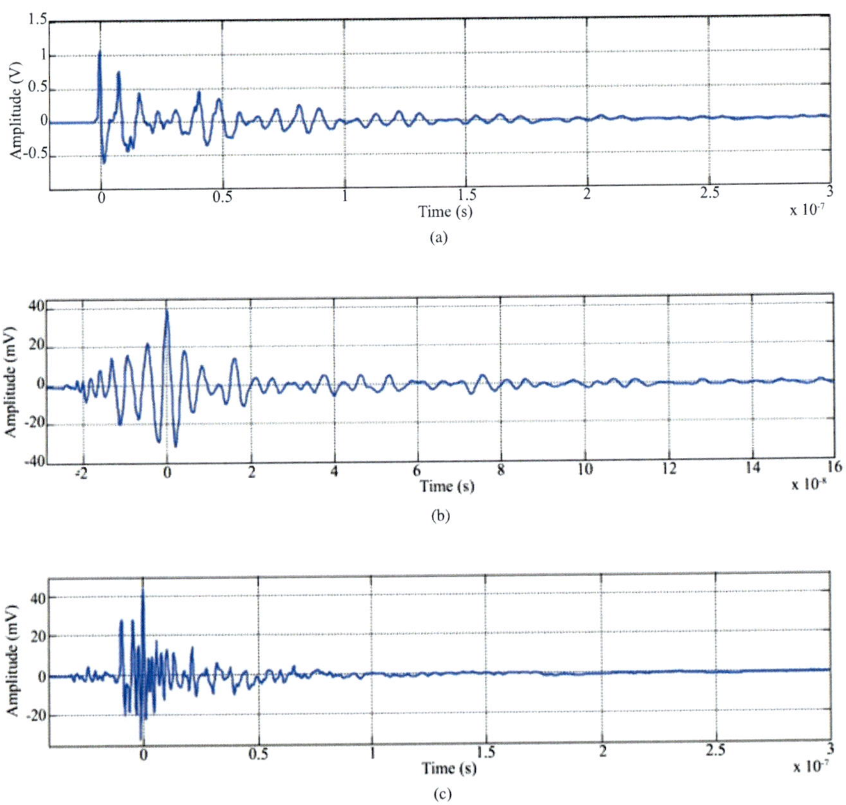

Fig. 3.17 Time domain signals spectra of internal delamination measured using: (**a**) directional coupler, (**b**) log-periodic antenna and (**c**) loop antenna

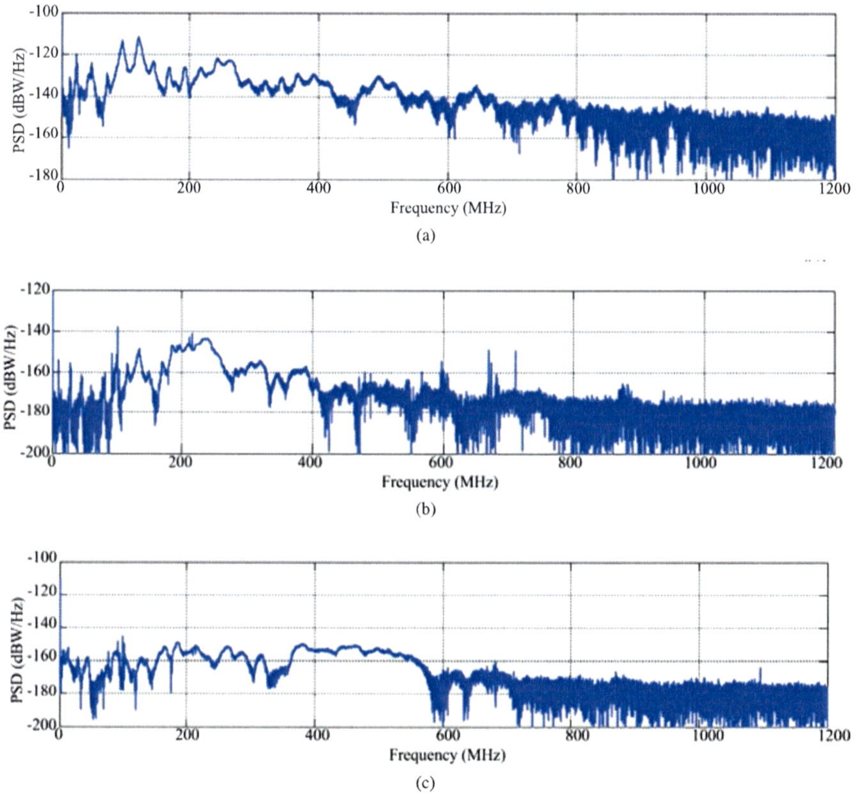

Fig. 3.18 Frequency spectra of internal delamination measured using: (**a**) directional coupler, (**b**) log-periodic antenna and (**c**) loop antenna

sensors, namely: a directional coupler, a log-periodic antenna and a loop antenna, respectively. These signals were obtained during an experiment involving PDs in an air-filled cavity of a power generator stator bar. It is possible to observe that the signal obtained using the directional coupler presents higher amplitude than the others. It is also clear that the frequency components of the signal obtained with the log-periodic antenna are more intense for frequencies under 400 MHz. Moreover, the signal obtained with the loop antenna has significant amplitudes up to 600 MHz similarly to the directional coupler.

The results of this experiment demonstrate that each sensor is sensitive to the PDs in a characteristic way. The directional coupler is able to capture the highest amplitudes, while the log-periodic antenna is best for capturing certain frequency components of the signal (between 180 MHz and 400 MHz, in our example). The loop antenna, although it presents the lower amplitudes, shows lower levels of spectral amplitude variations between the a few kHz to almost 600 MHz. It is also a good device to capture low-frequency oscillations at a distance. Thus, each sensor should be carefully chosen depending on the application.

3.4.1.3 Corona Activity at the S/C and Stress Grading Coating

Corona activity at the surface of the semiconductor coating (S/C) can be observed due to the presence of sharp edges, cracks and other imperfections. Those imperfections can cause high electric field gradients at the surface of the S/C, which can lead to PDs. To reduce the incidence of PDs, one of the most effective methods is the application of a stress grading coating. This coating is applied to the surface of the S/C and consists of a non-conductive material with a high impedance, such as polyurethane or epoxy resin. The purpose of this coating is to reduce the electric field gradients by acting as a buffer between the S/C and the environment. The stress grading coating can also be used to reduce the amount of noise produced by PDs.

This kind of PD activity occurs directly either at the junction of the S/C or at stress grading coating when the field grading system is not adequate, resulting in high local electrical stresses. This type of activity, simulated in laboratory during a voltage endurance test (VET), is illustrated by Fig. 3.19a and b. A typical PRPD pattern corresponding to this type of activity is shown in Fig. 3.20. Figure 3.21 presents the time domain signals and Fig. 3.22 shows the corresponding frequency spectra measured using: the directional coupler, the log-periodic antenna, and the loop antenna, respectively. It is observed that the three sensors register signals with a frequency spectrum mainly up to 500 MHz (we may consider that the loop antenna registered appreciable amplitudes up to 600 MHz).

Figure 3.19 shows the abrasion indicative in the transition zone between the semi-conductive and voltage stress control layers in different parts of the coil. In Fig. 3.19a, the abrasion is present in the lower area of the coil, while in Fig. 3.19b, it is in the upper area of the coil. This evidence suggests that PDs occurred in those areas of the coil. PDs can cause damage to electrical components due to the high temperature generated by the spark and the high voltage applied. The presence of such discharges in the transition zone between the semi-conductive and voltage stress control layers of the coil indicates that the insulation system of the coil has been compromised.

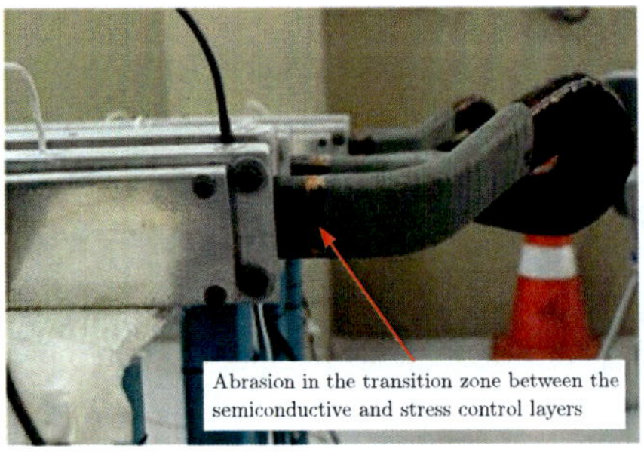

(a)

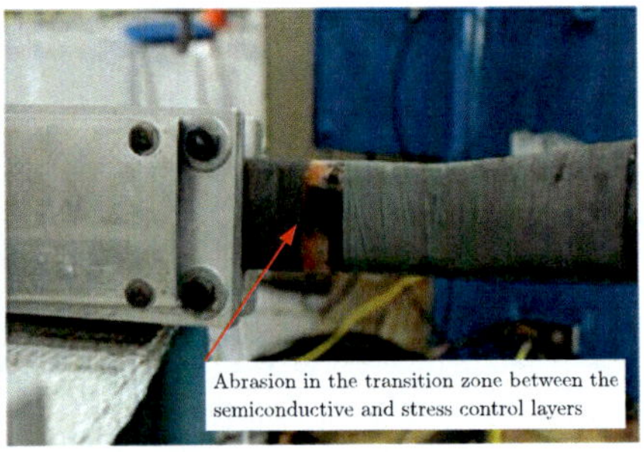

(b)

Fig. 3.19 Abrasion on transition zones between the semi-conductive and voltage stress control layers at: (**a**) the lower area of the coil and (**b**) the upper area of the same coil

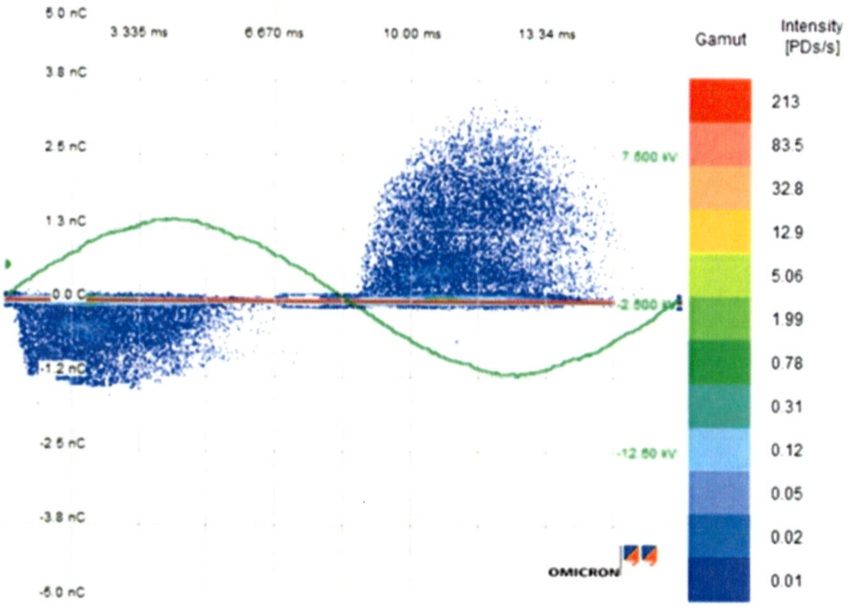

Fig. 3.20 Corona activity: typical PRPD map obtained in laboratory

Figure 3.20 shows a typical corona PRPD map obtained in our laboratory. With maps like that in Fig. 3.20, one can estimate the risk and severity of corona PD activity in insulation systems. It is known that PD activity increases with the electric field and, consequently, with the electric potential applied to the insulation surface [13]. Thus, it is expected that the PDs activity will be higher in the corners and edges of the insulation material, where the electric field is higher. The presence of PDs can cause partial or total insulation failures, depending on the magnitude of the PDs. For this reason, it is of great importance to monitor and control the PDs activity in insulation systems.

Figure 3.21 presents the signals in time domain and Fig. 3.22 corresponding frequency spectra measured using the directional coupler, the log-periodic antenna, and the loop antenna. These measurements were used to characterize the PDs generated in our laboratory. The directional coupler was used to measure the fast pulses of the PDs, while the log-periodic and loop antennas were used to measure the pulsed emission of PDs at distance. It was found that the PD signals had a wide frequency spectrum, ranging from kilohertz to megahertz. The results of this study demonstrate that the mentioned measuring techniques are suitable for characterizing PDs in laboratory conditions. Once more, we see that loop detected signals from a few kHz to almost 600 MHz, the directional coupler detected signals from a few kHz to 400 MHz and the log-periodic antenna shows PD energy between 180 MHz and 420 MHz.

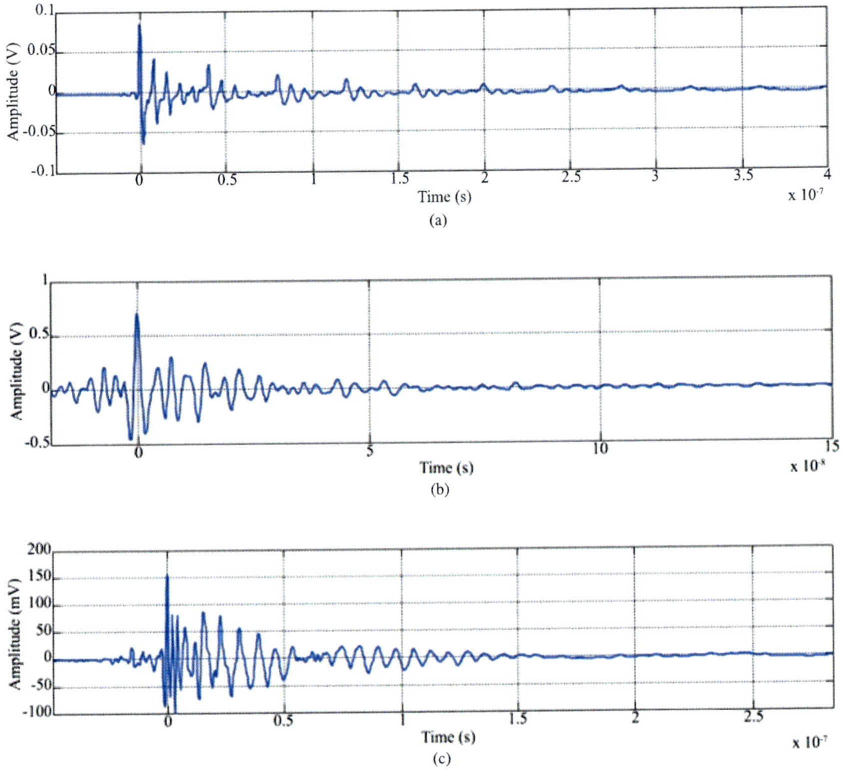

Fig. 3.21 Time domain signals spectra of corona measured using: (**a**) directional coupler, (**b**) log-periodic antenna and (**c**) loop antenna

3.4.1.4 Slot Discharges

Slot discharges are generated in air or gas-filled pockets within the stator core, between the surface of a bar and the stator core. This activity arises when the electrical contact between the S/C coating of the bar and slot is lost. In laboratory tests, this type of fault is illustrated in Fig. 3.23. A PRPD pattern recorded in our laboratory during slot discharges activity is shown in Fig. 3.24. Figure 3.25a and b demonstrate signals in time domain and Fig. 3.26a and b their corresponding frequency spectra obtained using the directional coupler and the log-periodic antenna respectively. Directional coupler detected PD power up to approximately 1.0 GHz, while the log-periodic antenna registered significant PD power up to about 750 MHz.

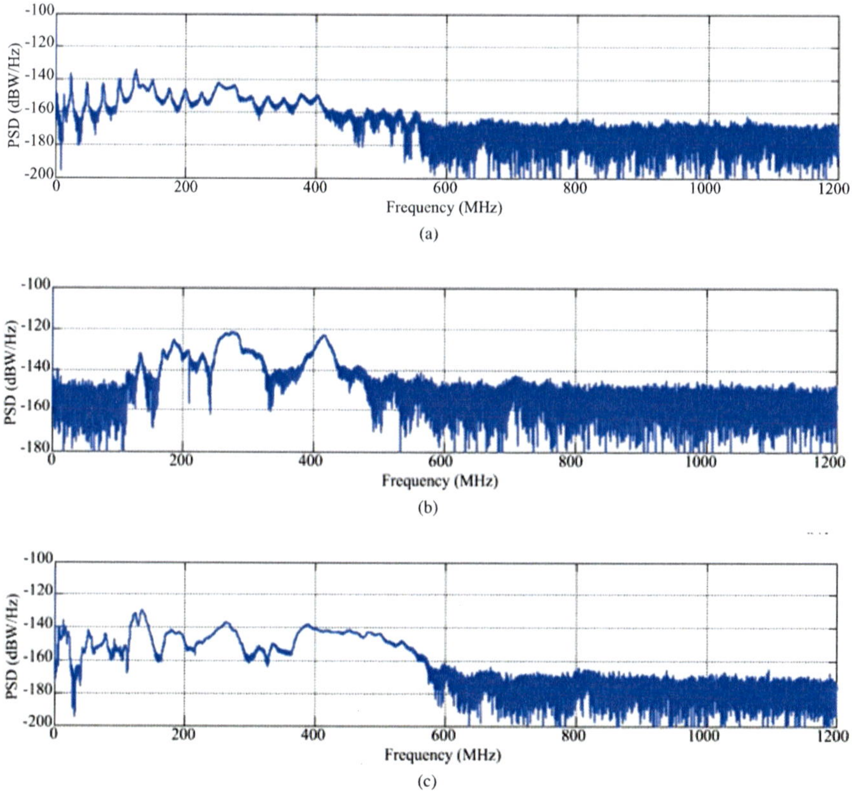

Fig. 3.22 Frequency spectra of corona measured using: (**a**) directional coupler, (**b**) log-periodic antenna and (**c**) loop antenna

3.4.1.5 Gap Discharges

This type of activity occurs between internal conductors of the bar or between a bar and the press finger of the stator core, as illustrated in Fig. 3.27. A typical PRPD pattern corresponding to this type of activity is shown in Fig. 3.28. Figure 3.29 present the measured signals in the time domain and Fig. 3.30 present the corresponding frequency domain signals for gap discharges, acquired with the directional coupler, the log-periodic antenna, and the loop antenna. The directional coupler and the loop antenna detected signals with significant spectral amplitudes up to 300 MHz. However, the log-periodic antenna was able to register appreciable amplitudes up to 400 MHz, but it was not able to detect low-frequency signals (under approximately 50 MHz).

Figure 3.27 shows gap faults, which is a type of PD frequently observed in electrical insulation systems. It is caused by the presence of a gap between two conductive parts, creating a fault in the insulation between them. It is possible that

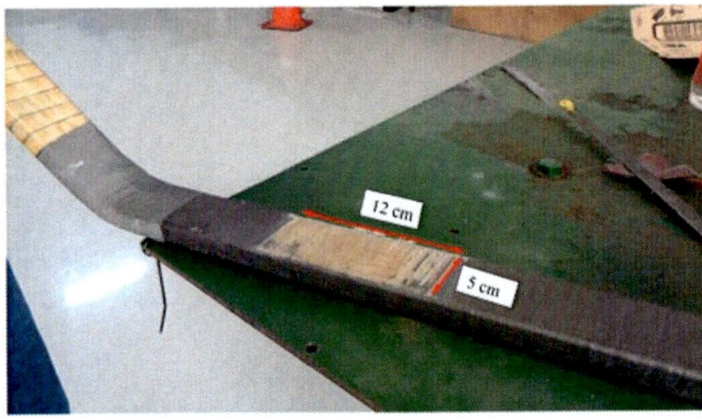

Fig. 3.23 Slot discharges: intentional abrasion of part of the semiconductor coating to simulate slot discharges

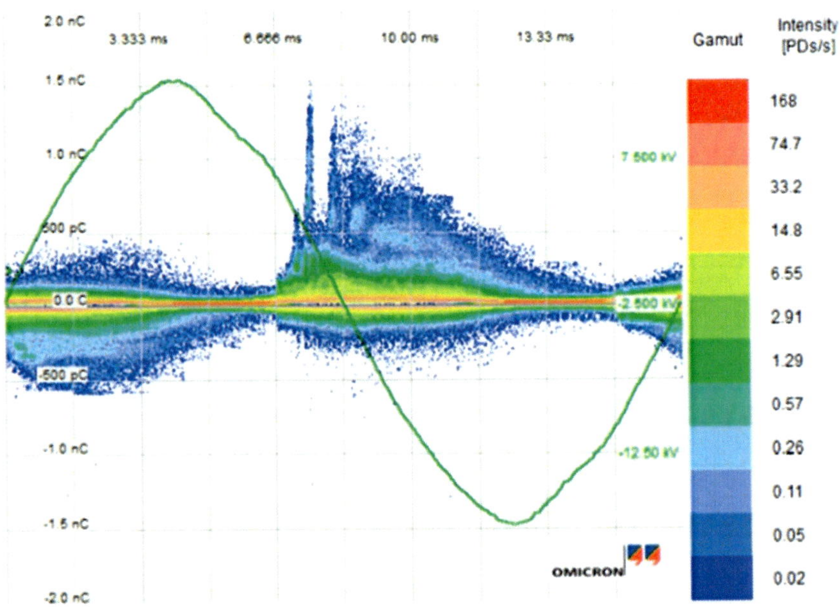

Fig. 3.24 Slot discharges: typical PRPD of slot discharges obtained in laboratory

kind of fault results in PDs that occur between the internal conductors of a bar, as depicted in Fig. 3.27b. Gap faults are usually caused by poor manufacturing, installation and maintenance practices, and they can lead to severe damage in electrical equipment. Therefore, it is essential to diagnose and assess these faults in order to prevent potential critical damages.

Gap faults can be replicated and studied in a laboratory using the setup shown in Fig. 3.27a. This setup consists of two pairs of conductors separated by an insulation material inside a hydrogenerator coil, which can be of different types depending on

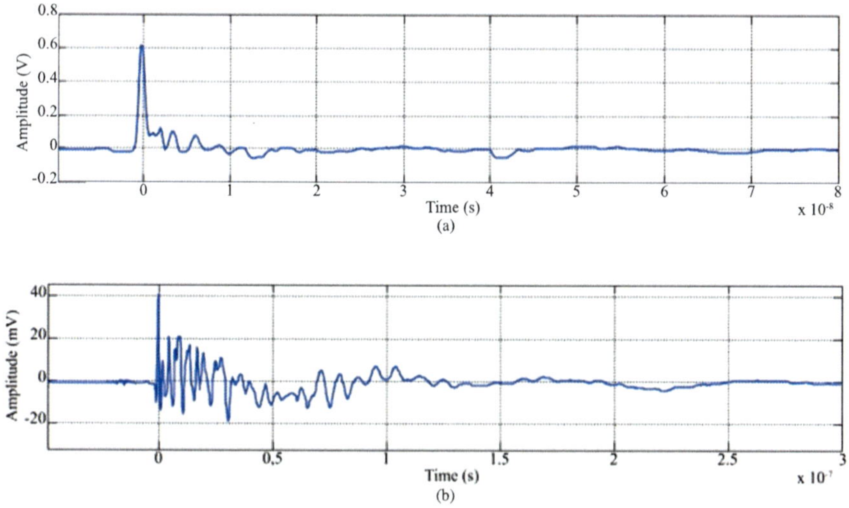

Fig. 3.25 Time domain signals spectra of slot measured using: (**a**) directional coupler and (**b**) log-periodic antenna

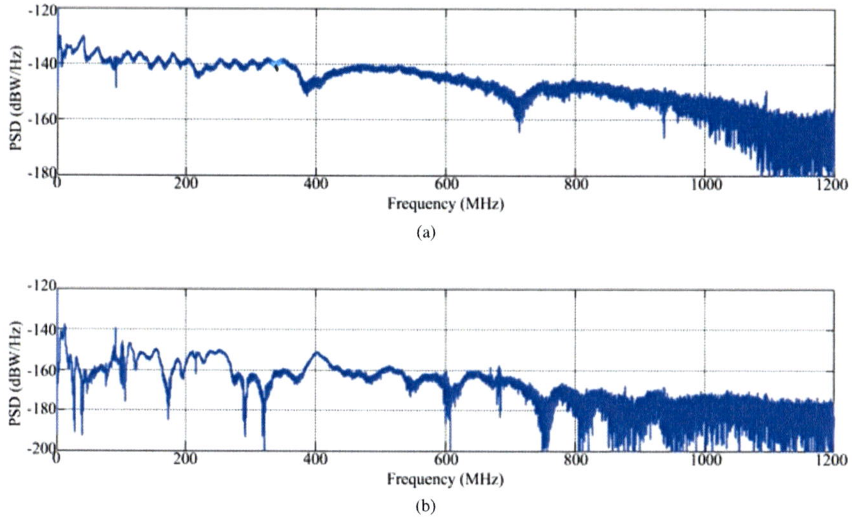

Fig. 3.26 Frequency spectra of slot measured using: (**a**) directional coupler and (**b**) log-periodic antenna

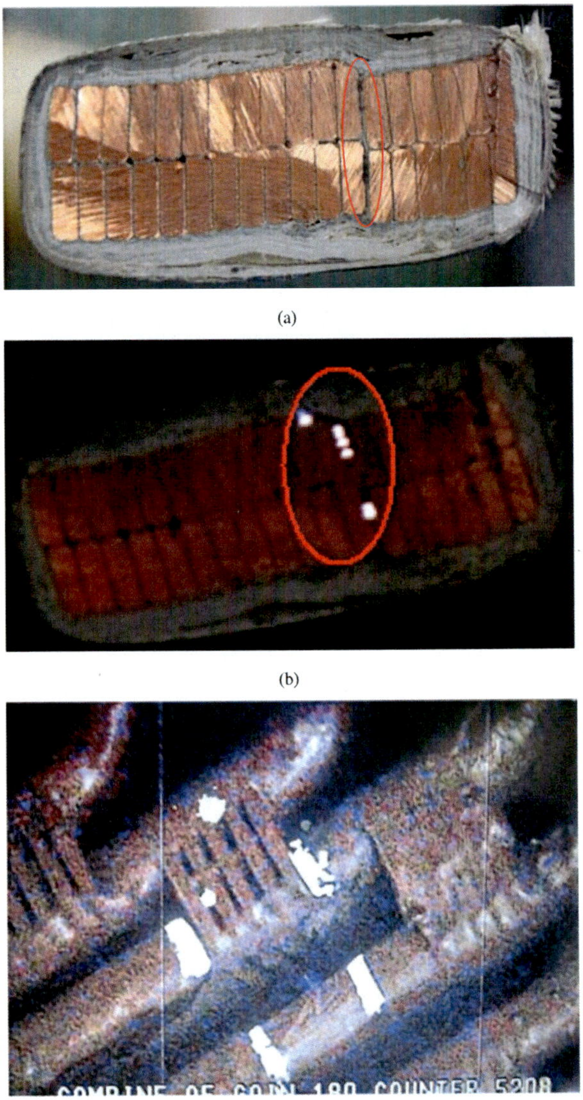

(a)

(b)

(c)

Fig. 3.27 Gap faults: (**a**) internal gap fault replicated in laboratory (forced fault in insulation amid the internal conductors), (**b**) PD occurrences among the internal conductors of a bar in laboratory and (**c**) external gap PDs in operating generator winding

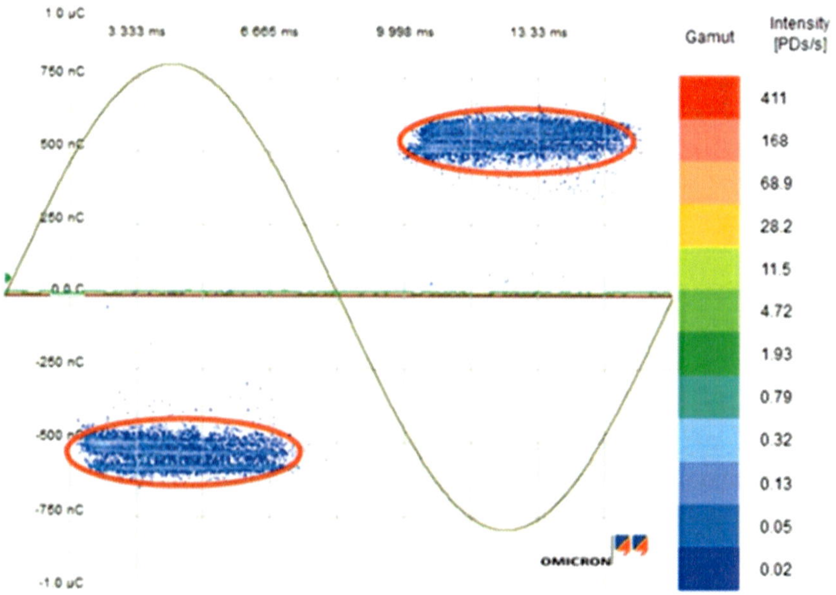

Fig. 3.28 The gap fault: typical PRPD of gap obtained in laboratory

the application and the insulation system used. The resulting fault creates a space between the two conductors, thus allowing the emergence of the PDs. Furthermore, the setup can be used to measure the characteristics of the PDs. This information can then be used to identify and assess the gap fault.

The magnitude of the electric field is also a function of the distance between the electrodes of the fault site [14]. Hence, the monitoring of PD under different electric fields can be done by changing the distance between the electrodes. As the distance between the electrodes is reduced, the electric field increases, leading to a growth in PD levels [14]. Figure 3.28 shows a typical PRPD map of gap PDs measured in laboratory. As we see, horizontal clouds are formed, making a distinguishable and unique PRPD pattern, which is easy to recognize. Similar PRPD maps are obtained for the case of Fig. 3.27c.

3.4.2 *Surface Discharges (Tracking)*

Surface discharges occur along the winding due to contamination of winding surfaces, on the interface between a high-voltage surface and the air. This problem was replicated in the laboratory, as illustrated in Fig. 3.31. The contamination of a coil surface was implemented by using a copper tape, which is used to excite

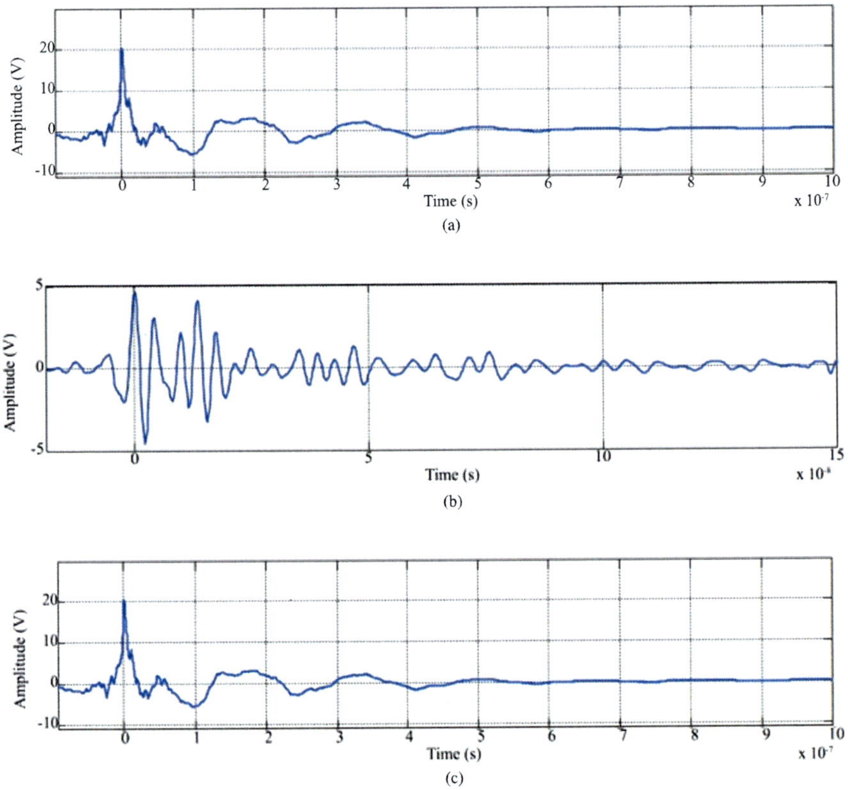

Fig. 3.29 Time domain signals spectra of gap measured using: (**a**) directional coupler, (**b**) log-periodic antenna and (**c**) loop antenna

tracking PDs. Copper tape works such as possible impurities on the insulation surface, leading to local high electric fields that can cause breakdown. Figure 3.32 shows a typical tracking PRPD pattern recorded in our laboratory.

Figure 3.33a, b and c show measured time domain signals and Fig. 3.34 present the respective frequency domain signals, measured by using the directional coupler and also the log-periodic and the loop antennas.

Usually, tracking PDs are phase-localized and with fluctuating amplitudes, producing vertical PRPD clouds. It is thus possible to use PRPD as an indicator of the severity of the contamination on the insulation surface, since tracking PRPD profile is easy to distinguish. This can be done by comparing the generator PRPDs with laboratory-obtained maps, such as that shown in Fig. 3.32. This comparison is important in order to detect the presence of PDs and determine their type and severity, thus helping to avoid possible damages to the insulation system.

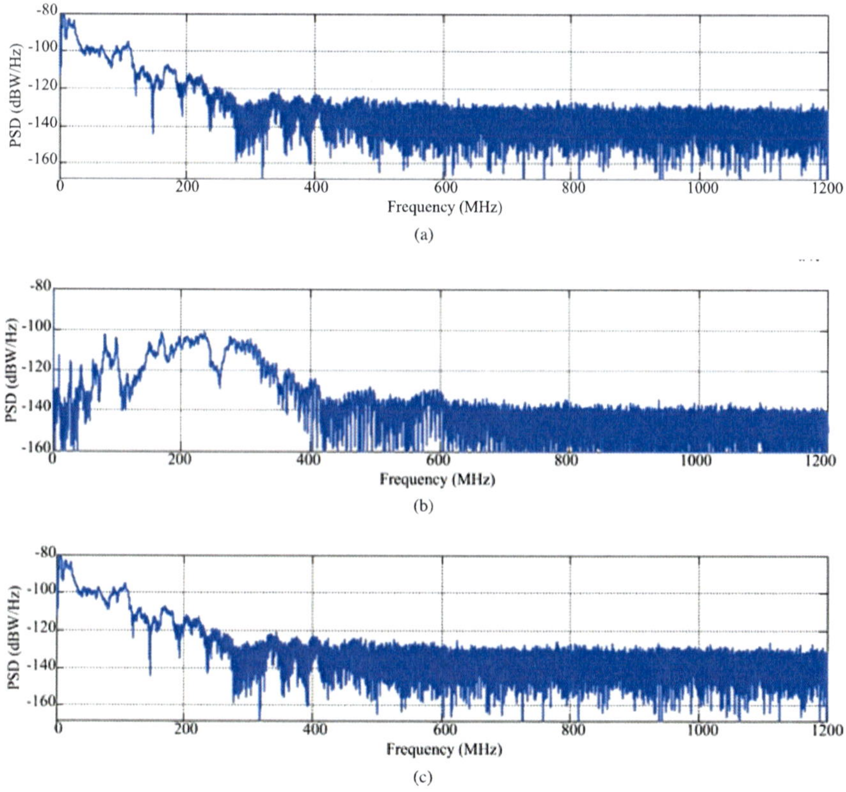

Fig. 3.30 Frequency spectra of gap measured using: (**a**) directional coupler, (**b**) log-periodic antenna and (**c**) loop antenna

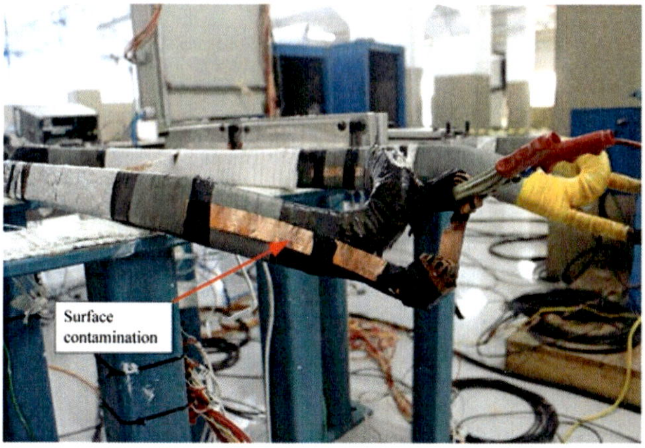

Fig. 3.31 Surface discharge: a copper tape used as contamination object on the coil surface

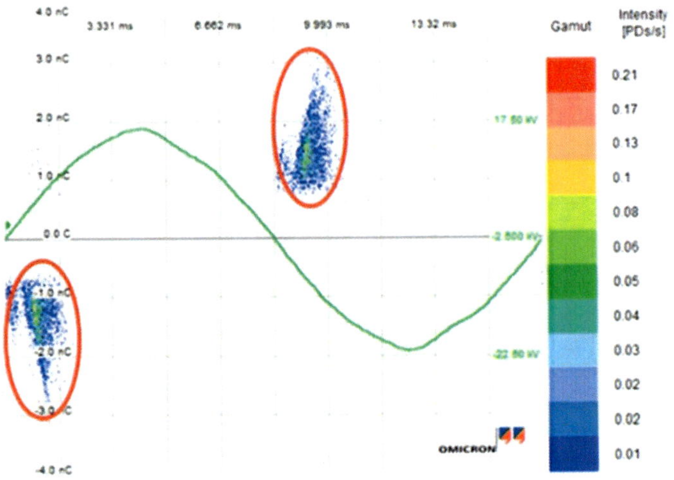

Fig. 3.32 Surface discharges: typical PRPD of surface contamination obtained in laboratory

Figure 3.33 show signals in the time domain and Fig. 3.34 displays the respective frequency domain signals, once more obtained by using the directional coupler, the log-periodic antenna and the loop antenna. The time-domain signals can be used to identify the magnitude and features of transient PD signals and to study repeatability of discharge pulse shapes over time, while the frequency-domain signals can be used to identify spectral peaks and characteristic spectral bands. Consequently, these signals can be used to diagnose the condition of the insulation.

Table 3.2 presents a summary of the frequency ranges recorded using the electromagnetic sensors for each type of discharge. It is observed a strong correlation between the type of fault and the frequency band of the electromagnetic signals radiated due to the corresponding PD activity.

3.5 Final Remarks

The results presented in this work show that it is possible to accurately characterize different types of PDs using the frequency components of the frequency spectrum in the signals captured using the electromagnetic sensors. The directional coupler presented the greatest amplitude sensitivity in comparison with the loop and log-periodic antennas. However, antennas can be used to detect PDs at distance, being alternatives as less intrusive measuring devices then directional couplers. Log-periodic antennas can provide high directionality and narrower bandwidth, while loop antenna provides lower directionality.

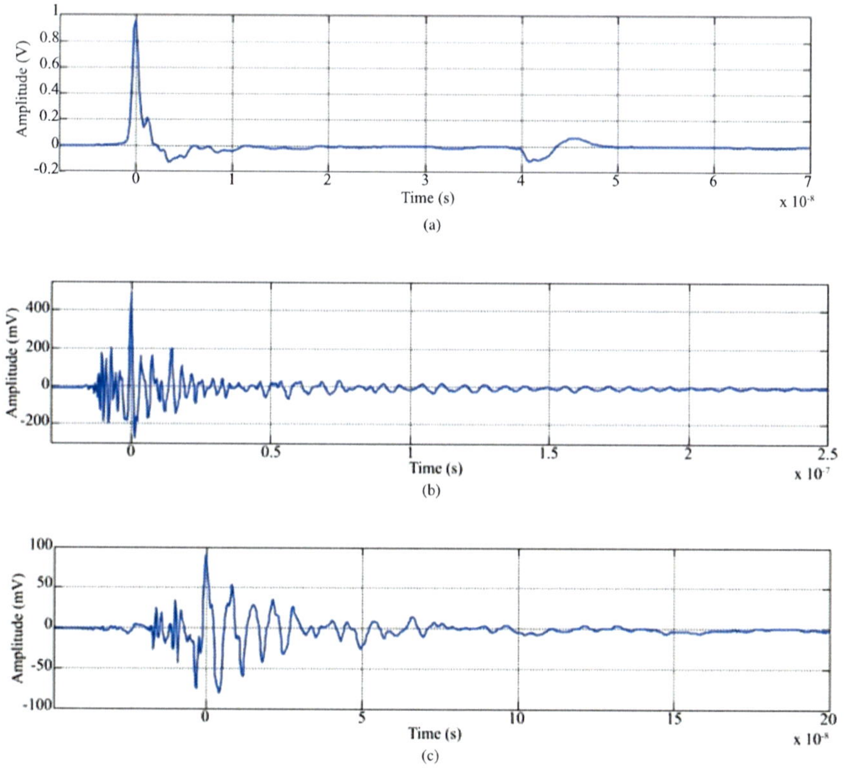

Fig. 3.33 Time domain signals spectra of surface PDs measured using: (**a**) directional coupler, (**b**) log-periodic antenna and (**c**) loop antenna

Finally, frequency band of each type of PD is provided. Internal voids range from a few kHz to 700 MHz, delamination from a few kHz to 800 MHz, corona activity at the surface/conductor and stress grading coating from a few kHz to 500 MHz, slot discharges from a few kHz to 1000 MHz, gap type discharges from a few kHz to 300 MHz, and surface discharges tracking from a few kHz to 900 MHz.

With the objective of developing a monitoring system for the hydrogenerator, more tests and investigations are necessary to analyze the behavior of the system under various working conditions. Certainly, new emerging techniques such as wavelet transforms and advanced artificial intelligence can be associated to new measuring techniques to provide each time more robust PD diagnostics.

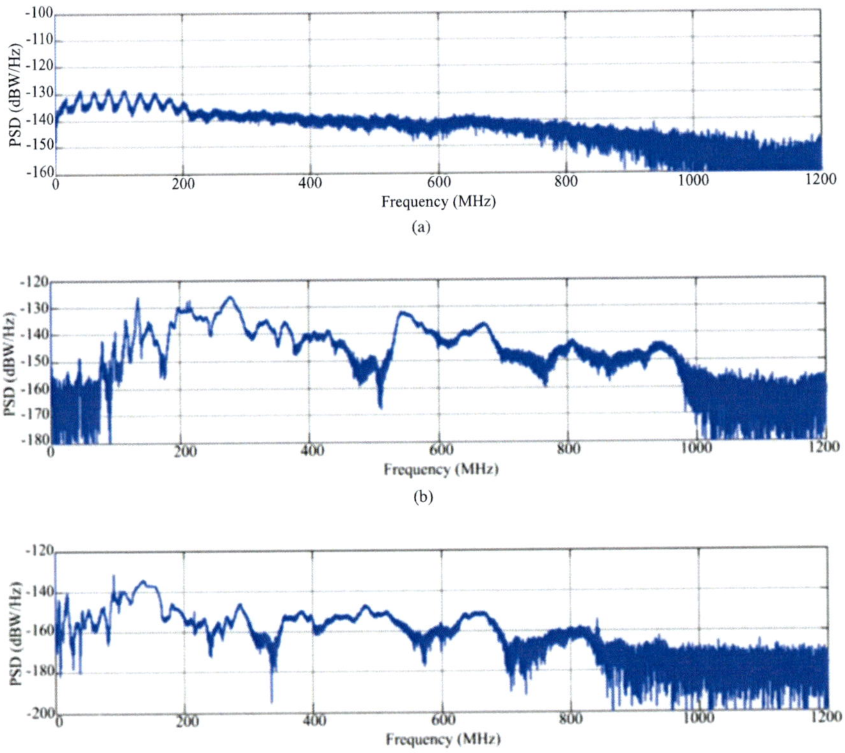

Fig. 3.34 Frequency spectra of surface PDs measured using: (**a**) directional coupler, (**b**) log-periodic antenna and (**c**) loop antenna

Table 3.2 Frequency range for each type of PD (≈ 0 means a few kHz)

Type of discharge	Frequency range (MHz)
Internal voids	≈ 0–700
Delamination	≈ 0–800
Corona activity at the S/C and stress grading coating	≈ 0–500
Slot discharges	≈ 0–1000
Gap type discharges	≈ 0–300
Surface discharges (tracking)	≈ 0–900

References

1. International Electrotechnical Commission. (2000). High-voltage test techniques-partial discharge measurements. IEC, Publication-60270.
2. Institute of Electrical and Electronics Engineers. (2014). IEEE guide for the measurement of partial discharges in AC electric machinery. IEEE Std 1434-2014 (Revision of IEEE Std 1434-2000) (pp. 1–89).

3. Rotating Electrical Machines - Part 27-2. (2012). On-line partial discharge measurements on the stator winding insulation of rotating electrical machines. IEC/TS 60034-27-2. International Electrotechnical Commission, Technical Report.
4. Institute of Electrical and Electronics Engineers. (1993). IEEE standard definitions of terms for antennas. IEEE Std 145-1993 (pp. 1–32).
5. Rozi, F., & Khayam, U. (2015). Development of loop antennas for partial discharge detection. *International Journal on Electrical Engineering and Informatics*, *7*, 29–41.
6. Rozi, F., & Khayam, U. (2014). Design, implementation and testing of triangle, circle, and square shaped loop antennas as partial discharge sensor. In *The 2nd IEEE conference on power engineering and renewable energy (ICPERE)* (pp. 273–276).
7. Sena, A. J. C., de Oliveira, R. M. S., & do Nascimento, J. A. S. (2021). Frequency resolved partial discharges based on spectral pulse counting. *Energies*, *14*(21).
8. Khayam, U., & Fatoni, F. I. (2017). Design and application of loop antenna for partial discharge induced electro- magnetic wave detection. In *2017 6th international conference on electrical engineering and informatics (ICEEI)* (pp. 1–6).
9. Balanis, C. A. (2016). *Antenna theory: Analysis and design* (4th ed.). Wiley.
10. Cavallinil, A., Montanaril, G. C., Candela, R., & Testa, L. (2008). Partial discharges detection in medium voltage systems using directional antennas sensors. In *Annual report conference on electrical insulation dielectric phenomena* (pp. 455–458).
11. Robles, G., Martínez-Tarifa, J. M., Rojas-Moreno, M. V., Albarracín, R., & Ardila-Rey, J. (2012).Antenna selection and frequency response study for UHF detection of partial discharges. In *2012 IEEE international instrumentation and measurement technology conference proceedings* (pp. 1496–1499).
12. Rohde & Schwarz (2020). R&S HL223 log-periodic antenna: Optimized for radiomonitoring and measurements. https://scdn.rohde-schwarz.com/ur/pws/dl_downloads/dl_common_library/dl_brochures_and_datasheets/pdf_1/HL223_2020.pdf.
13. Stone, G., Culbert, I., Boulter, E., & Dhirani, H. (2014). *Electrical insulation for rotating machines: design, evaluation, aging, testing, and repair* (2nd ed.). Wiley-IEEE Press.
14. Florkowski, M., Krześniak, D., Kuniewski, M., & Zydroń, P. (2020). Partial discharge imaging correlated with phase-resolved patterns in non-uniform electric fields with various dielectric barrier materials. *Energies*, *13*(11).

Chapter 4
Partial Discharge Measurements in Synchronous Generators

4.1 Types of PDs in Synchronous Generator

The stator coil of a rotating machine can exhibit PD activity at different levels, depending on the characteristics—location, severity, and materials—of isolation defects. Figure 4.1 shows a transverse cut of a stator bar with typical defects [1, 2]. The following comments refer to defect types pointed out in Fig. 4.1:

(a) Corona in coil heads is caused by intense electric field at conductor corners in the early stages of a machine operation. The sharpness of the conductor corner, which is an important element for the occurrence of PDs, depends on the manufacturing process;
(b) Conductor-insulator delamination is the principal insulation wear in a generator. The delamination of the insulator layers usually results from machine overloading;
(c) Insulation detachment inside delamination of the epoxy-mica insulation is a consequence of the process of thermal aging;
(d) Regions without mica layers are due to imperfect curing of the insulation system during manufacturing;
(e) Slot discharge is caused by vibration of the bar inside the slot, which in turn might be due to inadequate wedge or partial vibration of the core;
(f) Gas-filled cavities inside a solid insulation material result from imperfections of the manufacturing process.

Faults in stator insulation consist in low impedance paths between conductors—coils, bars, phases, and core—that should ideally be insulated from each other. Table 4.1 lists usual stator winding faults, their mechanisms and symptoms, detection test procedures, and commonly associated machine types [3]. Stator winding faults can be generated by several processes: voltage-dependent electrical stress, vibration, overheating, chemical attack, contamination, and humidity. Over time,

V. Dmitriev et al., *Partial Discharges in Hydroelectric Generators*, Power Systems,
https://doi.org/10.1007/978-3-031-36604-8_4

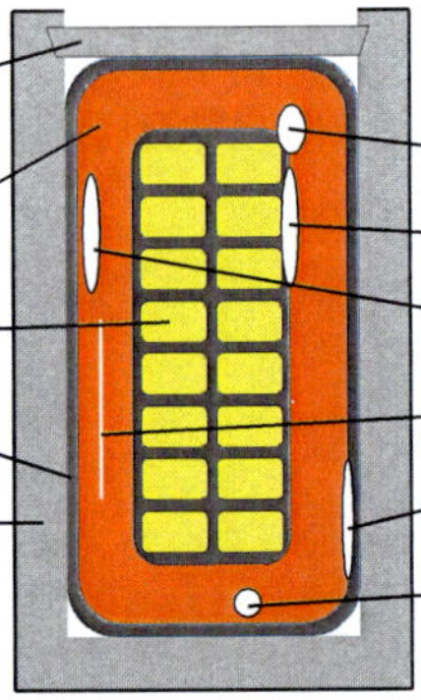

Fig. 4.1 Cross-section of a stator bar indicating structural elements (numbers) and common isolation defects (letters). The characteristics of a PD depend on the type of defect and its position in the structure. Source: adapted from [2]

such faults increase the deterioration of the insulation system, which is indicated by a greater presence of PDs.

4.2 Approaches and Systems for Measuring PDs

There are several methods to measure the PD levels during operation (i. e. online measurements) in motors and generators [5]. Electrical techniques require monitoring the current or voltage pulse that is created, whenever a PD occurs. The first methods for measuring PD pulses were based on the detection of high frequency currents in the machine neutral, but today most approaches use high voltage capacitors as sensors [6, 7] and the typical values of capacitance are 80, 220, and 500 pF.

Online PD measurements in motors, generators, and turbogenerators are particularly challenging because, once a machine is connected to the power system, electrical noise is always present and can mislead PD analysis. For instance, one can erroneously conclude that the stator windings exhibit high PD levels, when it is just a noise. For this reason, effective noise reduction procedures are needed to avoid false alarms, whose recurrence undermines the credibility of online PD tests as a reliable tool for monitoring condition of high voltage equipment [8]. Common noise sources include corona from the power system, ring collector, commutator, ignition and sparks from poor electrical connections and power system operation.

Assuming denoising procedures are successfully performed, features from recorded PD pulses e.g., pulse counting and phase positions can then be reliably determined. The data collected in a typical evaluation session are usually summarized by diagrams or merit figures. Among them, one has a number of positive and negative peaks of PD amplitudes with minimum occurrence rate of 10

Table 4.1 Stator winding faults: failure mechanism, symptoms, tests, and associated machine types

Failure mechanism	Symptoms	Detection test	Machine types
Inadequate impregnation	PD	PD measurement, tan δ [4], and power factor	Vacuum impregnation and resin tape
Failure in semiconductive layer	PD, slot discharge, and ozone	PD measurement, visual inspection, and ozone monitoring	Air-cooled
Loose winding	PD, ozone, and loose wedges	PD measurement, visual inspection, ozone monitoring, and wedging tests	Systems with thermosetting insulation (epoxy and polyester)
Sparking by vibration	PD, ozone, and loose wedges	PD measurement, visual inspection, ozone monitoring, and wedging tests	Insulation based on resin tape or vacuum impregnation
Strain relief interface	PD, white powder (nitric acid), and ozone	PD measurement, tan δ, power factor, visual inspection, and ozone monitoring	Air-cooled and with painting in the strain relief interface
Inappropriate spacing	PD, white powder, insulation discoloration, and ozone	PD measurement, visual inspection, and ozone monitoring	Air-cooled, circuit-breaker, and motor connections
Coil head cooling	Loss of blocking and binding system, dispersed white powder or grease	Visual inspection, fiber optic accelerometer	High voltage, high velocity with great coil heads
Improper electrical connection	PD, white powder and connection discoloration	PD measurement, visual inspection, and measurement of infrared temperature	Any connection in all types of machines
Thermal deterioration	PD, insulation discoloration	PD measurement, tan δ, power factor, and visual inspection	All types of machines
Load cycles	PD, insulation failure in the bar ending	PD measurement, tan δ, power factor, visual inspection, surge test, and Hipot (high potential or high voltage)	All types of machines (with long bars)

pulses per second (+Qm and −Qm, respectively). The parameter Qm is considered a reasonable predictor of the stator insulation condition. The higher values of Qm, the more deteriorated are the windings [8]. Another example is the phase-resolved partial discharge (PRPD) graph (Fig. 4.2), where the vertical axis corresponds to PD amplitudes, while the phase position is given by the horizontal axis. Information about the frequency of occurrence of such amplitude-phase pairs is coded in color.

The destructive nature of PD in the insulation system of high voltage devices has been known for more than a century [8]. Hence, it is not surprising that PD measuring methods have been developed since 1930. A list of applications that rely on PD measurements is as follows [9]:

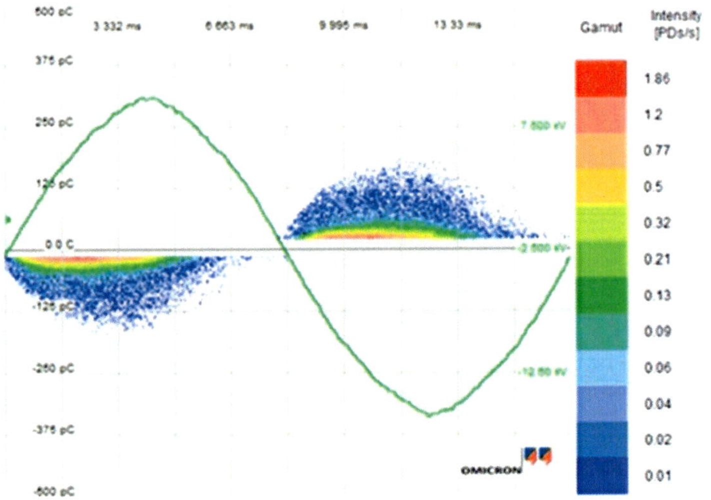

Fig. 4.2 Example of a PRPD pattern. The occurrence of amplitude-phase pairs is coded in color; the vertical axis gives PD amplitudes, the horizontal axis corresponds to phase position

(a) design test: evaluation of a new insulation system to ensure that PD activity under normal operation is absent or below a specified level;
(b) quality assurance test: checking whether voids were introduced during fabrication or processing of the insulation system;
(c) diagnostic test: determination whether electrical insulation has not been deteriorated during operation as a result of electrical, thermal, mechanical or environmental stress.

The purpose of these tests is to assess the insulation system integrity which is closely related to the condition of high voltage equipment. In other words, PD activity above certain levels raises a warning flag for the risk of machine failure. Over time, several methods for PD detection and data visualization have been developed, usually targeting improved strategies for quality assurance testing or design. Nowadays, some of such methods are usual in laboratories and industries [10–12] to prevent failure in high voltage equipment.

Characteristics of PD Measurement System

According to [13], a PD pulse has an extremely rapid rise time and a short width. The oscillation period, rise time, and peak magnitudes vary significantly for each PD pulse. These characteristics usually depend on the geometry of the machine, location of the pulses and insulation material.

PD pulses have a frequency spectrum ranging from DC to GHz. The bandwidth of a measuring equipment is limited and cannot characterize the whole PD phenomenon. There is a direct relation between device bandwidth and price, a key issue to take into account when planning a program for PD monitoring. Most

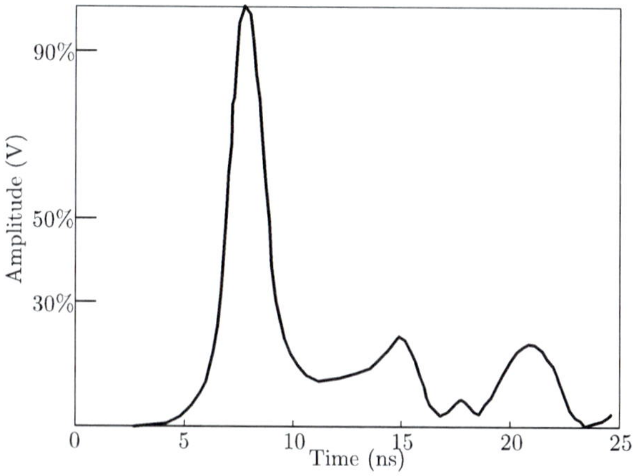

Fig. 4.3 Typical partial discharge pulse. Source: [10]

measuring instruments can only detect the onset of a PD pulse, which has a rise time mostly varying from 1 to 5 ns. We can use the rise time to approximate the pulse central frequency as follows:

$$f = \frac{1}{4 \times t_r}, \tag{4.1}$$

where f represents the pulse central frequency and t_r is pulse rise time. For example, for pulse with rise time of 3 ns, we estimate the central frequency $f = 1/(12$ ns$) = 83$ MHz. Thus, a rise time interval of 1 to 5 ns corresponds to a central frequency range of 50–250 MHz. Figure 4.3 depicts a typical PD pulse.

Nowadays, PD measurements are performed digitally. The simplest way to measure and display PD pulses is to use a digital oscilloscope, whose accuracy outperforms that of its analog counterparts. In Fig. 4.4, we see a PD pulse recorded by a portable digital oscilloscope [14] from a hydrogenerator at Coaracy Nunes power plant (Brazil).

A significant change in the area of PD measurement was promoted by the widespread adoption of digital recording using pulse magnitude analyzers (PMAs). Introduced to the PD community in the late 1960s [15], the PMA does not work as a regular oscilloscope, but separates positive from negative pulses, determines the value of and counts PD peak amplitudes. Moreover, PMAs provide an indication of PD repeat rate, while the oscilloscopes do not. The results of a PMA are usually displayed graphically, as seen in Fig. 4.5. Virtually all PD commercial detectors made in the past decade use PMA.

The PRPD analysis also known as PPA (pulse phase analysis), has an additional feature with regard to PMA, i.e. the phase angle of the AC cycle is also digitally recorded for each PD pulse. The result is a digital dataset rich enough to allow statistical inference about the type and severity of insulation defects. The PPA output is a two- or three-dimensional graph of the pulse count rate as a function of pulse magnitude and AC phase position. There are many ways to display this output, one

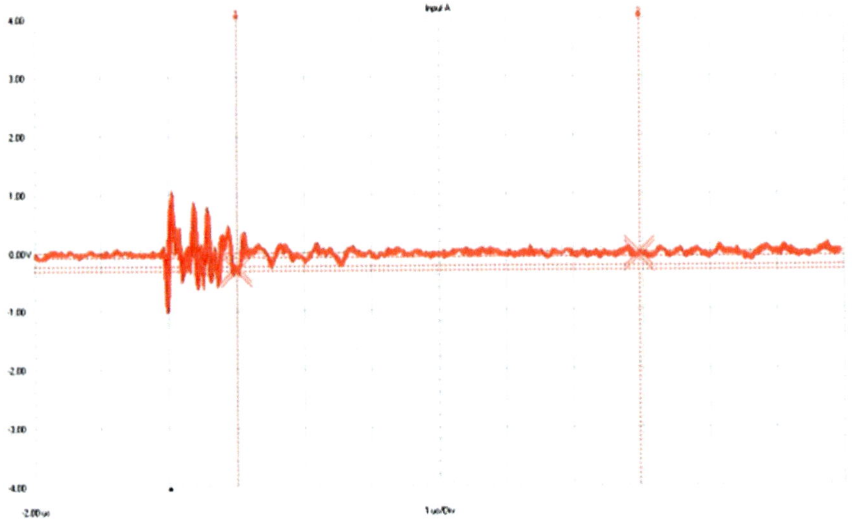

Fig. 4.4 PD pulse signal measured by a portable digital oscilloscope [14] in a hydrogenerator at Coaracy Nunes power plant (Brazil)

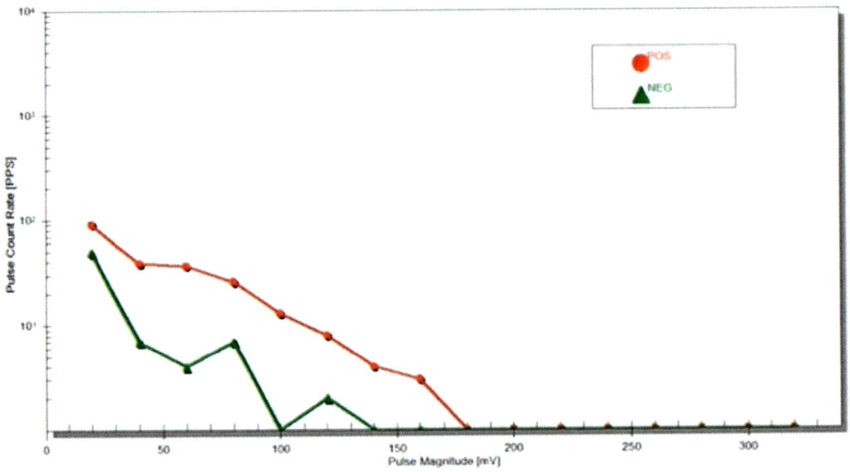

Fig. 4.5 An example of pulse magnitude analysis of partial discharges. Horizontal axis represents PD pulse magnitude and vertical axis shows the pulse count ratio in pulses per second. Color distinguishes positive (red) from negative (green) amplitudes

of which is shown in Fig. 4.6. Typically, the pulse repetition rate is displayed using a color code.

Digital instrumentation that produces PMA and PPA charts has reduced the need of experts for conducting PD tests. With analog technology, the presence of an expert during a PD test was mandatory to determine, for instance, pulse magnitudes

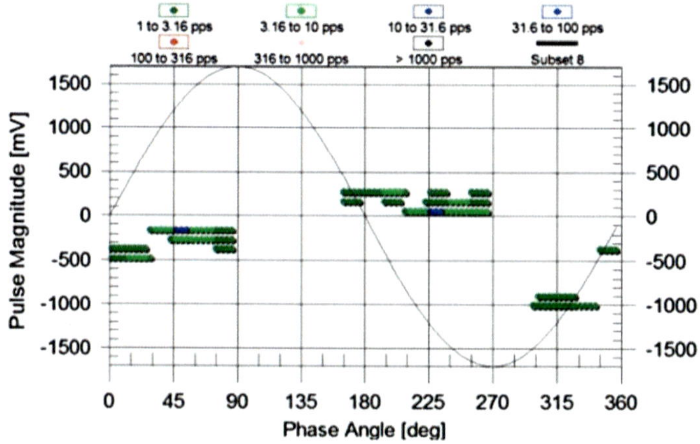

Fig. 4.6 An example of pulse phase analysis, also known as phase-resolved partial discharge analysis. Source: [9]

from the blinking strokes on the oscilloscope screen or to evaluate the level of noise in the measurements. With digital instrumentation, however, the expert does not need to be in site during measurement procedures and may review the data remotely, reducing the cost of testing. Most commercial PD instrumentation in use today employ analog-to-digital conversion with sampling rates about 20 MHz. These are fully compliant with IEC 60270 (narrowband and wideband detectors). Some suppliers, however, provide instruments with bandwidths up to 350 MHz to allow ultra-wideband PD detection with the consequent advantages of higher capabilities of noise separation and location of PD sources [9]. Measurements of PDs in the stator winding can be divided into two large categories: off-line measurements and on-line measurements. In off-line measurements, the stator winding is isolated from the power system and an independent voltage source is used to energize the winding. In online measurements, the machine operates regularly connected to the power system while PD measurements are made.

Important differences between on-line and off-line measurements are related to the voltage distribution along the winding as well as various thermal and mechanical effects associated with the machine operation, such as vibration, and temperature gradients between stator copper and iron core. Another example is the difference of the hydrogen pressure between off-line and online PD measurement in hydrogen-cooled machines [1].

4.3 Measurements in the Field

PD monitoring aims at keeping track of the production of new PD types and of the change in PD intensity levels. The identification of the PD type can be achieved by using PRPD charts. Correct PD type identification is vital, as each PD type is associated with a certain degree of risk, an aspect well established by the IEC/TS 60034-27-2 standard [1]. Table 4.2 presents a list with the type of PD, its typical PRPD pattern and the associated level of risk. Correlations between PD activity and other parameters of the machine—vibration, power and temperature—should also be taken into consideration. Each measurement session is intended to update the PD history and produce a complete technical report with all the information retrieved from recorded discharges.

Next, we report some results from measurement campaigns performed at the Brazilian hydroelectric power plants of Coaracy Nunes (Amapá State), Tucuruí (Pará State) and Samuel (RondÔnia State). The PD monitoring system used in all power plants, named *Instrumentation for Monitoring and Analysis of PD* (IMA-DP) [16], was developed by the Electrical Energy Research Center (CEPEL) and Eletrobrás Eletronorte in partnership. Figure 4.7 presents a diagram depicting a general PD monitoring scheme based on the IMA-DP system.

4.3.1 *Case 1: Coaracy Nunes Hydroelectric Generating Plant, Brazil*

In December 2009, a measurement campaign was carried out to evaluate the PD activity in machine 03 of Coaracy Nunes HPP (hydroelectric power plant), Amapá state, Brazil. We verified that PD signals in phase C distinctly differ from other phases. At that point, determination of the defect severity was not possible because of the unavailability of previous PD data from that machine. Then, the maintenance team decided to regularly measure PD signals from that time on to be able to draw a trend curve. Figure 4.8 shows one of the PRPD maps recorded on that occasion, whose pattern is due to surface discharges (see again Table 4.2).

During the inspection of the machine stator coil, some points of deterioration were identified between the front and bottom bars, in the lower part of the magnetic core (Fig. 4.9). Such a deterioration was probably a consequence of both corona effect and PD occurrence.

The problems identified by that inspection motivated a demanding maintenance intervention—the rotor needed to be removed—comprised of the following steps:

1. Cleaning the damaged area using a cloth with perchlorethylene or a similar solvent;
2. Removal of the insulating varnish around the damage without removing the anti-corona paint;
3. Sanding the damaged region without removing the anti-corona paint;
4. Cleaning the area with a dry cloth;
5. Painting the insulation and anti-corona protection with anti-corona paint;
6. Painting the damaged region with the insulating varnish after curing the anti-corona paint.

Table 4.2 Classification of discharges regarding type and degree of risk

Discharge type	PRPD pattern	Risk
Slot discharge	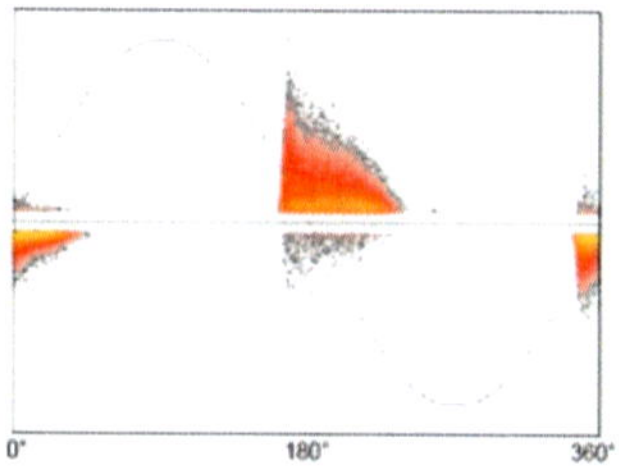	High
Delamination	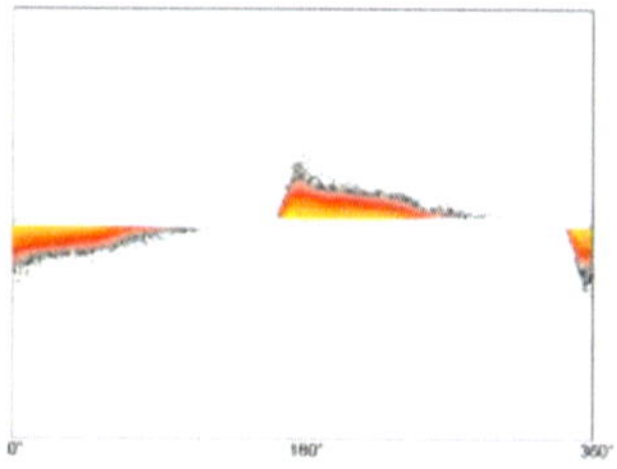	High
Delamination between conductor and insulation	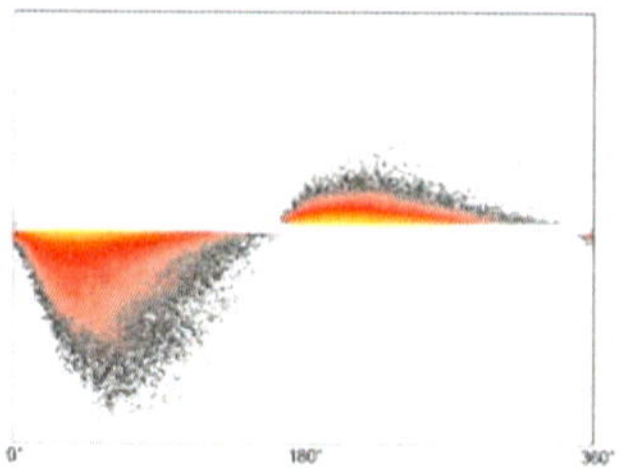	High
Gap discharge	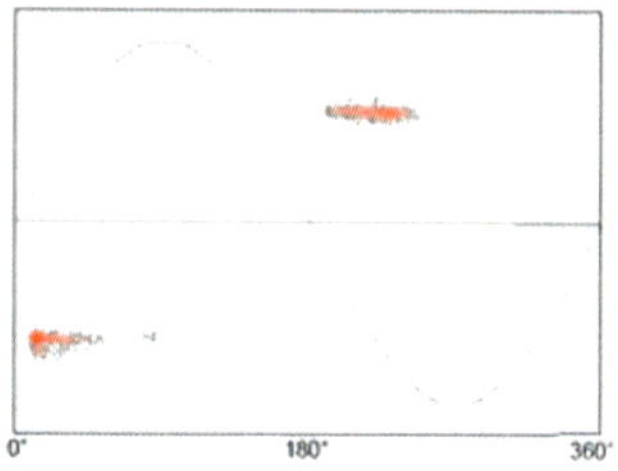	Medium
Surface discharge	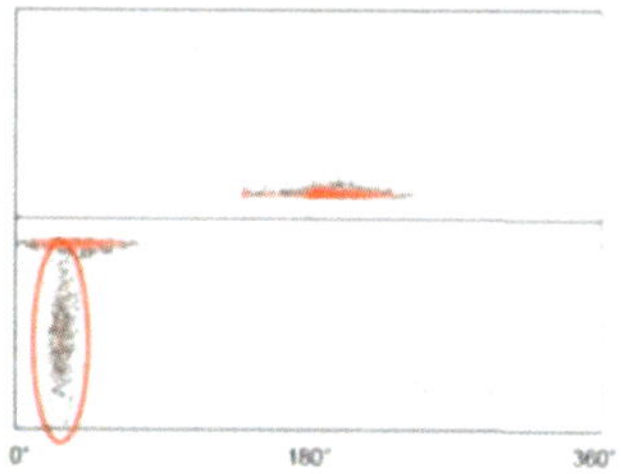	Medium

(continued)

Table 4.2 (continued)

Discharge type	PRPD pattern	Risk
Internal voids	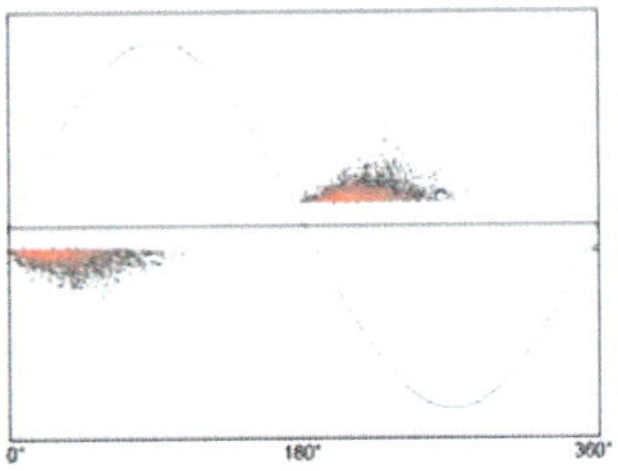	Low

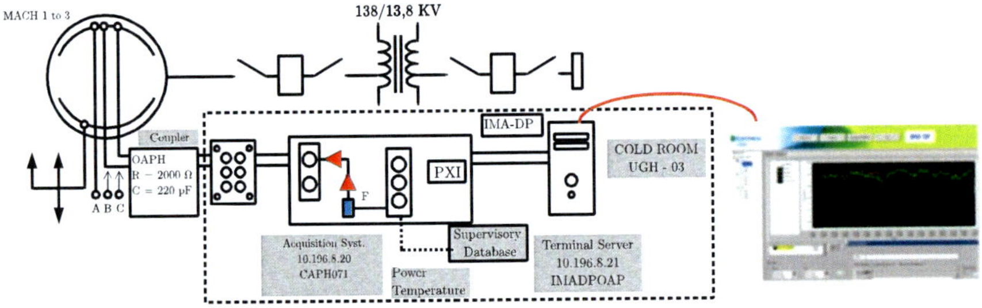

Fig. 4.7 Diagram showing a PD monitoring system based on IMA-DP. Capacitive couplers installed permanently in the output bars of the generator were used to monitor partial discharges. A three-channel oscilloscope is used to register transient signals derived from partial discharges. The PRPD measurements are accessed from database through an internal network

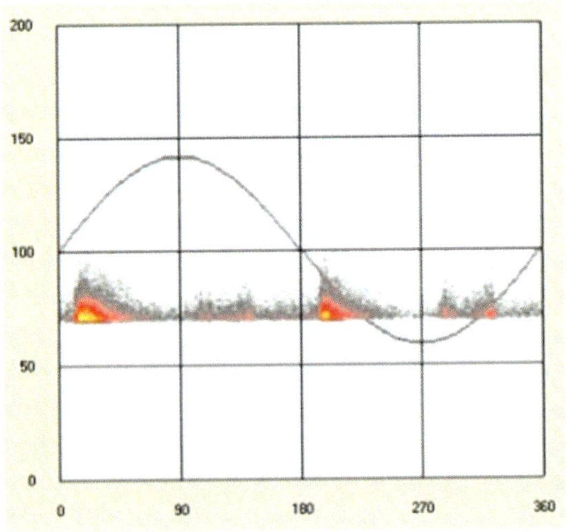

Fig. 4.8 Example of a PRPD pattern obtained from hydrogenerator 03 (CNUGH-03) at Coaracy Nunes power plant

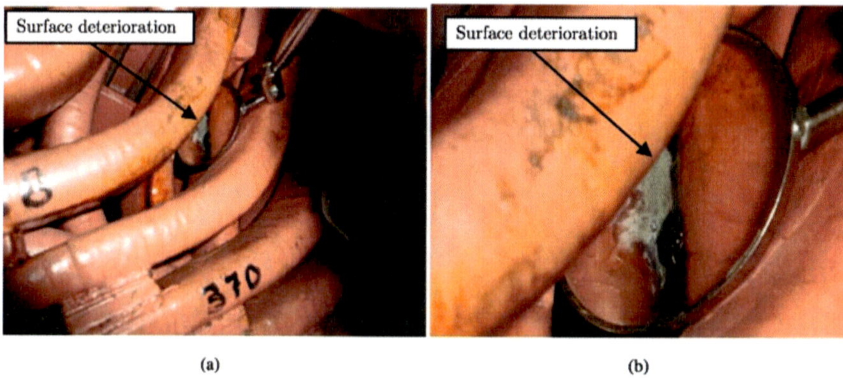

Fig. 4.9 Degradation of isolation system found during inspection of the stator coil (Coaracy Nunes power plant): (**a**) slot of bar number 214 exhibits resin impregnation leakage, possibly due to deficiencies in the curing process; also the bottom bar shows surface deterioration signs caused by corona effect or PDs; (**b**) zoom on the bottom bar

Fig. 4.10 Resulting PRPD after corrective intervention in the stator winding (Coaracy Nunes power plant)

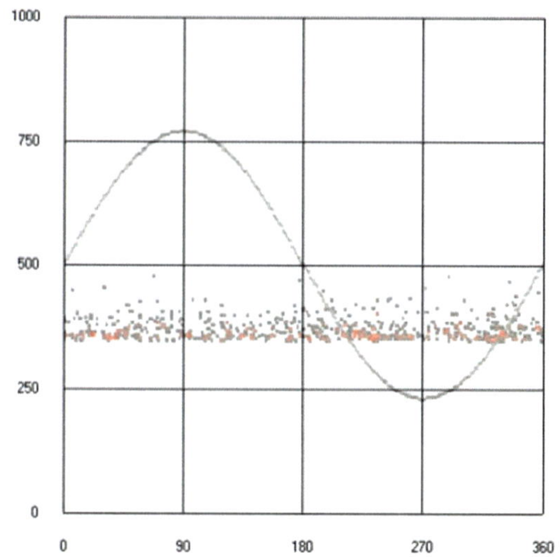

Figure 4.10 shows the resulting PRPD after the corrective intervention in the stator windings. Note the presence of a lower PD activity when compared with the PRPD before the intervention (Fig. 4.8). Another aspect to highlight is the absence of a marked PRPD pattern.

4.3.2 *Case 2: Coaracy Nunes Hydroelectric Generating Plant, Brazil*

Another measurement campaign in Coaracy Nunes HPP in Amapá state, Brazil was made in 2014. The three phases of each generator at the Coaracy Nunes HPP were monitored over the course of that year. In the months of continuous monitoring, the supervised machines operated properly, without exhibiting great increase in their PD activity. Observations showed, however, that the increases in the generated power are accompanied by a significant rise in the PD activity. Other machine operational conditions—like mechanical vibration and temperature changes—also demonstrated strong correlation with PD levels.

A. *Measured discharge patterns*
   Figures 4.11 and 4.12 give two examples where a prototype PRPD pattern in IEC/TS 60034-27-2 [1] is compared with the PRPD data found by IMA-DP. Note the measured PRPD needs an expert to be interpreted correctly, as actual PRPDs are generally formed by an association of simultaneous and distinct PD types.
B. *Correlation between PD activity and other variables (power, mechanical vibrations, and temperature)*
   This study was conducted on 30 June 2014 in a generating unit at Coaracy Nunes HPP; we first noted an increase around 100% between the PD levels measured at 3 p.m. and 10 p.m. Further observation of the operational conditions suggests the increase in PD levels was caused by a rise of the generated power by 3.75% above the rated power of the machine, which changed from 24 MW to 24.9 MW (Fig. 4.13). Additional analysis also reveals a 19% increase in relative vibration of the generator guide bearing (Fig. 4.14). The increase of PD activity also coincides with the rise in stator temperature (increase of approximately 3%) as can be seen in Fig. 4.15.

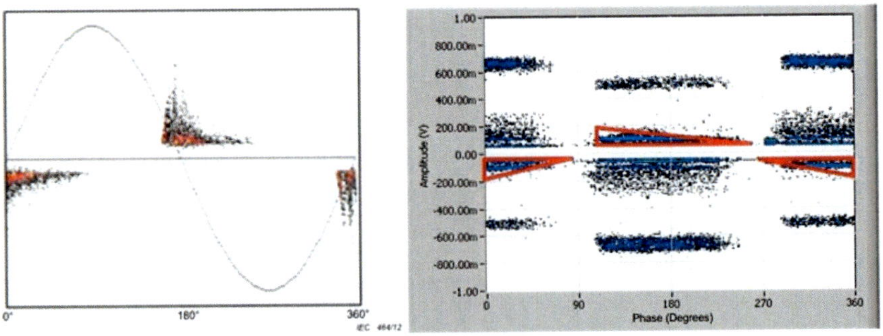

Fig. 4.11 Internal discharge due to internal delamination: (**a**) typical PRPD pattern presented in the IEC/TS 60034-27-2 standard [1]; (**b**) PRPD pattern acquired by IMA-DP at Coaracy Nunes power plant (the annotation in red indicates portion of the PRPD that is due to delamination)

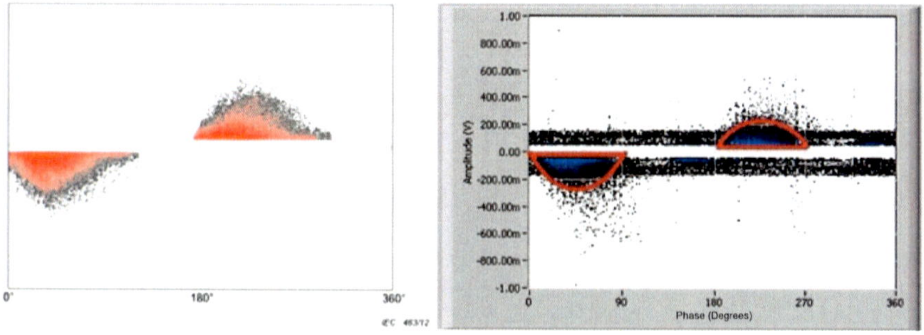

(a) (b)

Fig. 4.12 Internal discharge caused by internal voids: (**a**) typical PRPD pattern presented in the IEC/TS 60034-27-2 standard [1]; (**b**) PRPD pattern acquired by IMA-DP at Coaracy Nunes power plant (the annotation in red indicates portion of the PRPD that is due to internal voids)

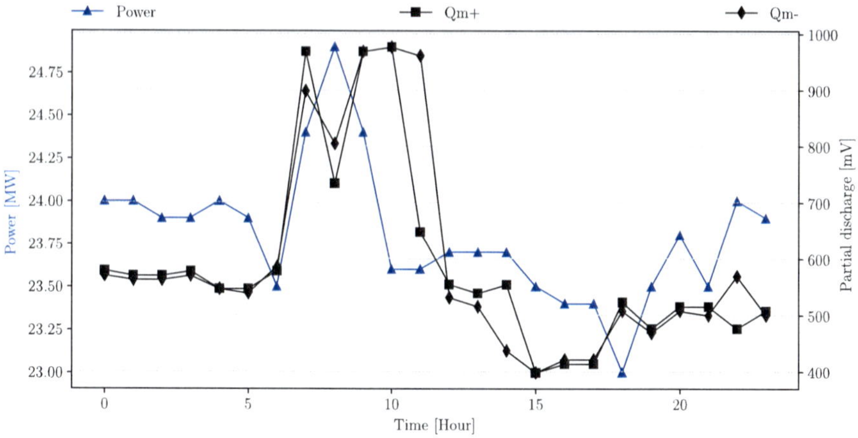

Fig. 4.13 PD magnitude (Qm) tendency and generated power by the machine 2 of Coaracy Nunes HPP

Even though we are aware that practical situations—like a measurement campaign in actual hydrogenerators—are not adequate for a rigorous study on how and to what extent mechanical vibrations, machine power and temperature interact to influence PD activity, our results indicate that the operating context must be taken into account when monitoring PDs to reach an exact diagnosis of the machine condition.

4.3.3 Case 3: Tucuruí Hydroelectric Generating Plant, Brazil

The following example was observed in one of the machines at the Tucuruí HPP, Tucuruí state, Brazil. In March 2016, during periodic measurement procedures, an

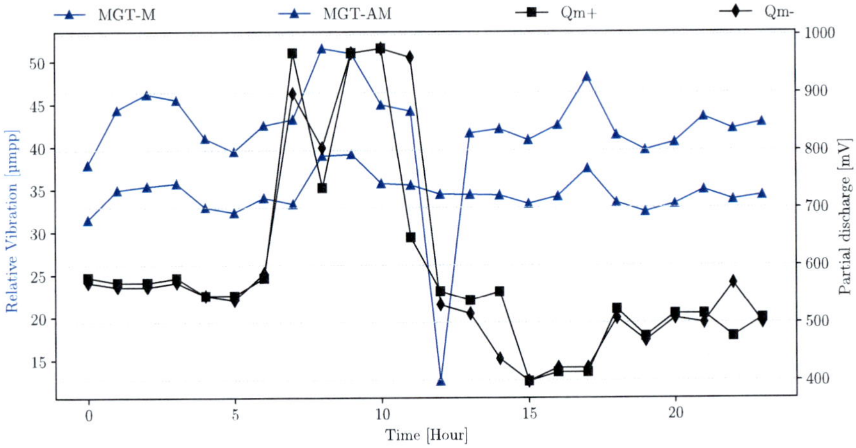

Fig. 4.14 PD magnitude (Qm) tendency and relative vibration in the turbine guide bearing (MGT) from the machine 2 of Coaracy Nunes HPP

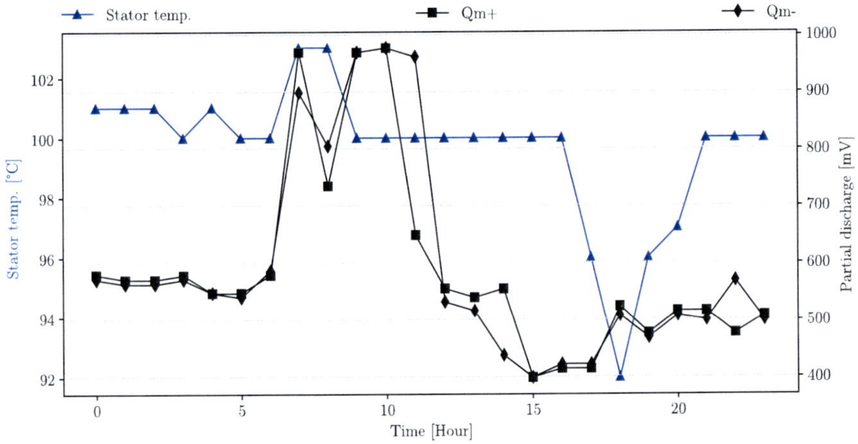

Fig. 4.15 PD magnitude (Qm) tendency and stator temperature over time from the machine 2 of Coaracy Nunes HPP

increase in PD levels of several couplers was observed, as can be seen in Fig. 4.16. PRPD patterns indicated surface discharges with high intensity (Fig. 4.17). After the intervention, PD patterns changed back to normal PD activity (Figs. 4.16, 4.17, and 4.18).

4.3.4 Case 4: Samuel Hydroelectric Generating Plant, Brazil

At Samuel HPP, located in Porto Velho state, Brazil we have an autonomous PD monitoring system based on IMA-DP, where measurements are made periodically

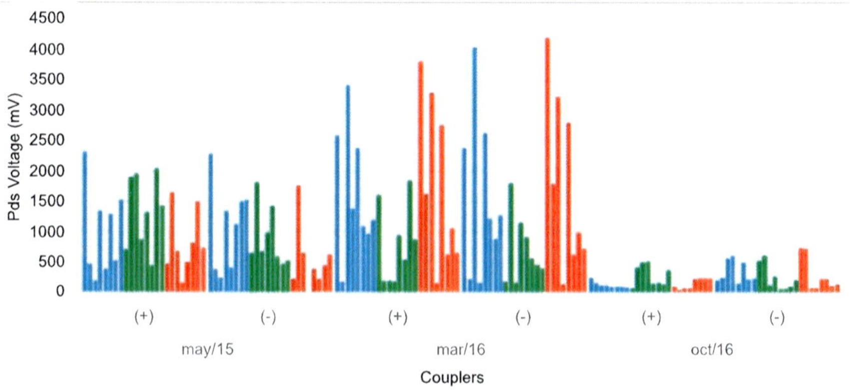

Fig. 4.16 PD levels recorded during regular monitoring procedures. Significant increase in PD activity verified in March 2016 motivated a corrective intervention in the concerned machine. In October 2016, after maintenance intervention, PD activity returned to normal levels

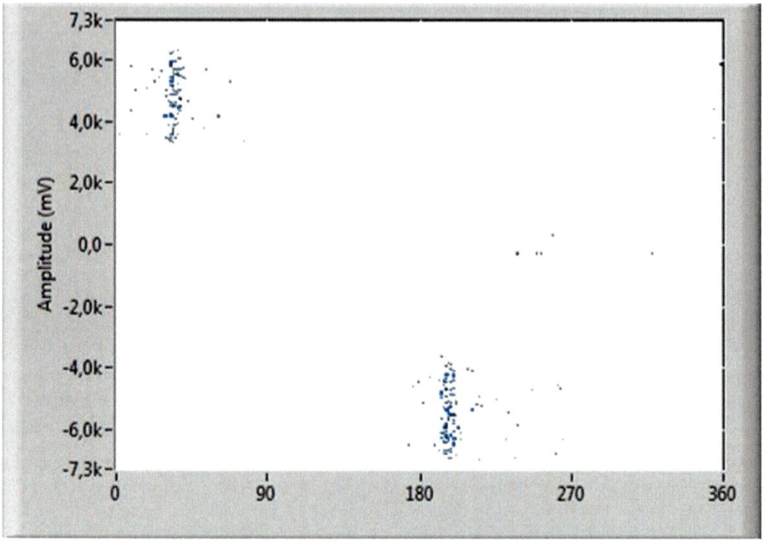

Fig. 4.17 PRPD pattern, obtained before intervention from the machine that exhibited the increase of PD levels, indicating the presence of surface discharges with high intensities

every 3 months. Samuel HPP is composed of five generating units of 50 MVA each. In September 2018, PD measurements revealed one of its units exhibited abnormal PD levels. Figure 4.19 presents the data collected in Samuel HPP, showing a considerable increase in PD levels for the concerned machine. For this reason, the interval between measurements was shortened to better track the PD evolution until the time defined to shut down the unit for inspection and maintenance. During maintenance in the assembly of the air guides, a metallic object was found. On

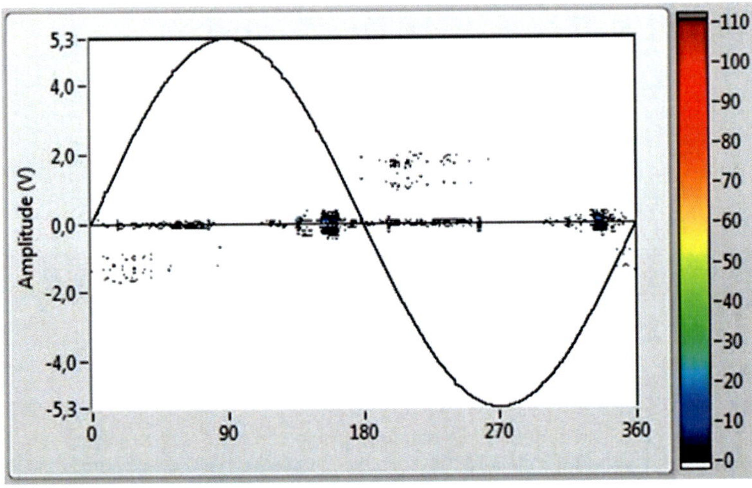

Fig. 4.18 PRPD pattern obtained after intervention, showing a return to normal PD activity

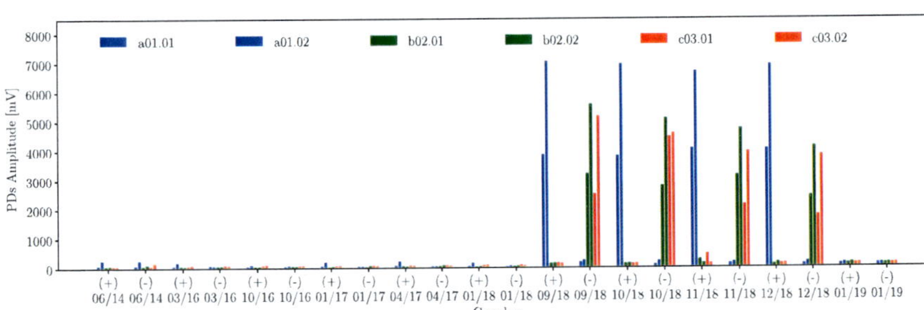

Fig. 4.19 PD levels recorded during regular monitoring. Significant increase in PD activity verified in September 2018 motivated a corrective intervention in the concerned machine. Capacitive couplers for phase A, B, and C are indicated by a01.0x, b02.0x, and c03.0x, respectively

site, the procedures of cleaning, sanding, cleaning again with isopropyl alcohol, and varnishing the affected area were carried out. After maintenance, the PD levels reduced to normal (see the last two measurements in Fig. 4.19).

4.4 PD Measurements During Commissioning Tests of a New Generator

Now we discuss PD measuring performed under the specific context of a commissioning test, where several variables, subjected to different operation conditions, are monitored to verify if the new generator meets its design specifications [17]. The main difference from a regular PD measurement is the opportunity to observe the interaction between a greater variety of thermal and mechanical parameters—vibration and temperature gradients for copper and iron core stators, for instance—and the PD activity [18].

4.4.1 Description of the Monitoring System

The IEC/TS 60034-27-2 standard [1], when describing online PD monitoring, emphasizes the importance of considering machine operating factors such as temperature, bar vibrations, and degree of contamination. Our next results were obtained from commissioning tests of a Francis turbine and a 611.1 MW power generator at Belo Monte HPP, a power plant built on the Xingu River, at 65 km out of Altamira, a city in Pará State (Brazil). For PD measurements, capacitive couplers of 1000 pF and the IMA-DP system were used. Figures 4.20 and 4.21 show capacitive couplers and IMA-DP installed at Belo Monte HPP premises, respectively [16].

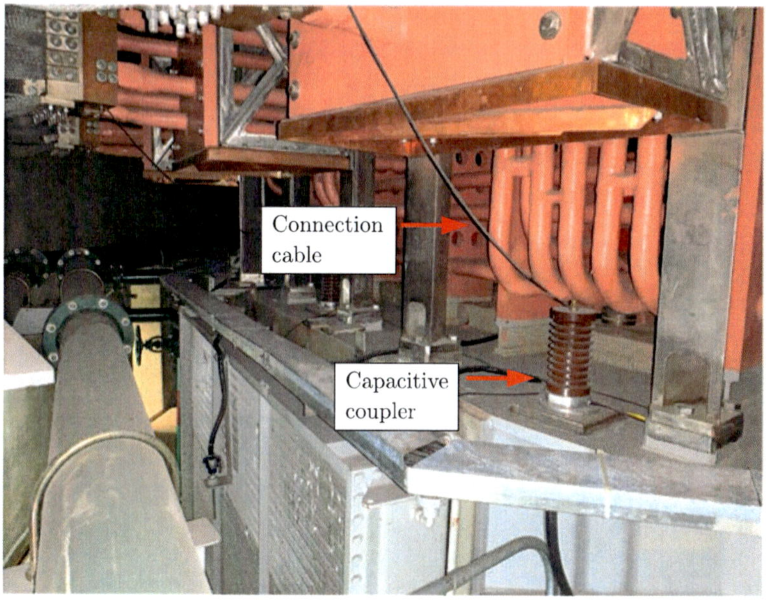

Fig. 4.20 Capacitive coupler of 1000 pF installed on the generator output bus and its connection cable

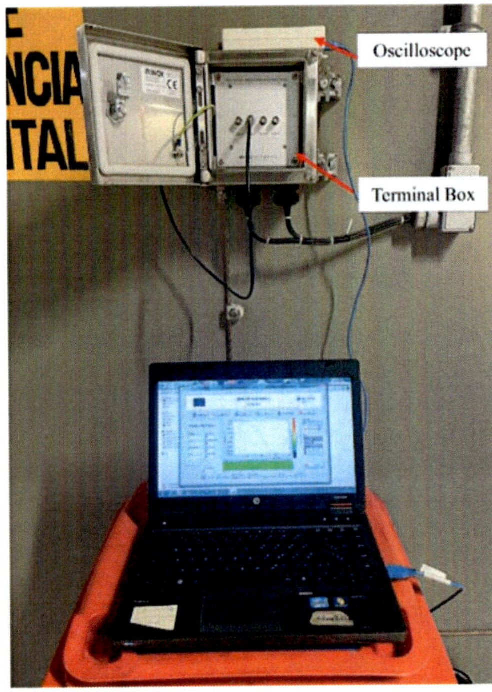

Fig. 4.21 IMA-DP system running on a laptop connected to an oscilloscope at Belo Monte HPP. The coaxial (black) cable connects the terminal box to the oscilloscope

4.4.2 Operating Curve Test

The operating curve test expresses the PD magnitude and the relative vibration as a function of the active power, while the generating unit is running with a load ramp ranging from 0 to 611 MW (i. e., from 0% to 100% of the rated power). Once the unit was synchronized, new measurements of both PD activity and relative vibration were made for each operating point. The latter is defined as the value of the active power, which varied from 0% to 100% of the rated power, with increments of 10% and an average stabilization time of 20 min at each level.

For active power between 10% and 40% of the nominal power (61.65 and 245.74 MW, respectively), the relative vibration reached a value that approximately corresponds to 40% of the vibration at rated power. In general, we observed that the lowest PD levels were achieved at nominal power, with the exception of phase A that presents PD levels under at nominal power for operating points of 50% and 60%. Figure 4.22a, b, and c present comparative graphs between relative vibration and PD activity for phases A, B, and C, respectively. Note the complex association between PD magnitudes and the relative vibration measured in the guide bearings of generator and turbine.

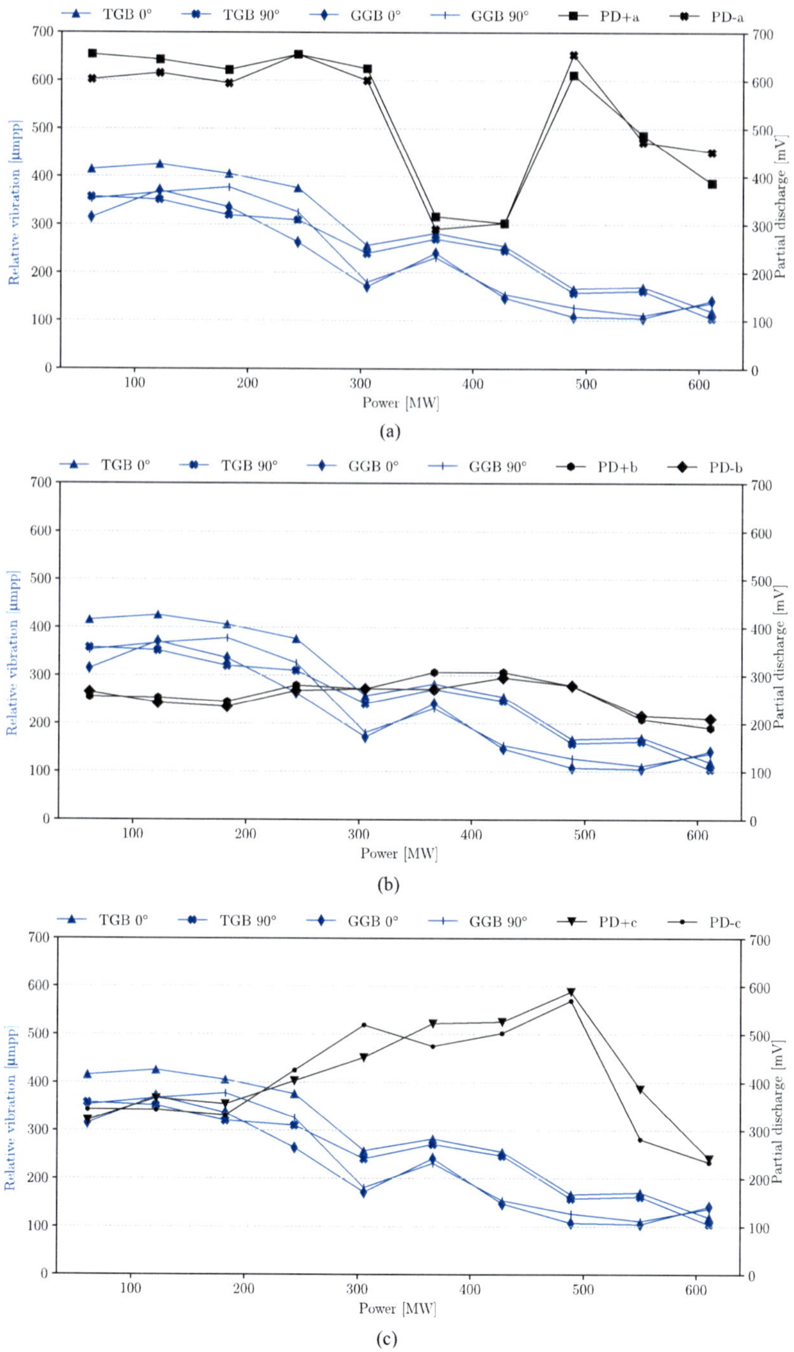

Fig. 4.22 Partial discharge (PD+, PD−) and relative vibration (TGB, GGB) as a function of the active power. TGB and GGB stand for turbine guide bearing and generator guide bearing, respectively. (**a**) Phase A. (**b**) Phase B. (**c**) Phase C

4.4.3 Generator Heating Test

The heating test is performed in the interval between 75% and 100% of the machine nominal load, following the method proposed in [19]. The information of interest is the rise of machine temperature for different loading conditions. Among several available loading methods we may use to obtain data, we chose the one that adopts a power factor (pf) larger than 0.9. Continuous loading tests should be performed until the machine temperature becomes constant—defined as a variation less than ±2 °C—for three consecutive half-hour readings. If the coolant temperature is not constant, the test may be terminated on the condition that the temperature rise does not exceed the maximum previously observed rises, taken into account at least three consecutive half-hour readings. If the coolant temperature for three half-hour readings varies by more than 2 °C, the test continues.

In the heating test, the following machine parameters are considered: active power (in MW), reactive power (in MVAr), power factor, armature voltage (phase) (in kV), armature current (in kA), field current (in A), field voltage (in V), and calculated rotor temperature (in °C). The temperatures of stator winding, stator core, heat exchanger, bearing metal, bearing oil, generator guide bearing metals, generator guide bearing oil, turbine guide bearing metal, and turbine guide bearing oil offer a considerable amount of data to compare with PD measurements.

In the generator heating test with 100% of the load, the temperatures of the stator and bearings stabilized[1] after 4.5 hours. Concerning PDs, positive and negative amplitude levels remained almost constant (between 200 and 300 mV) during the test, for a rated active power of 611 MW and a power factor of 0.9. During the test, the active and reactive power remained almost constant. Figure 4.23 compares PD levels to active and reactive power, armature and field voltages, and armature and field currents. Table 4.3 shows the variation of PD amplitudes from the beginning to the end of the heating test. The maximum variation was observed in phase C (PD+c). An important reduction of the PD levels was observed in tests with lower machine load (75% of nominal load) after stabilization of bearing temperatures.

Figure 4.24 presents the temperatures of rotor, stator, stator core, and heat exchanger, as a function of time. PD level curves are again displayed to facilitate comparisons. We note that the rotor temperature increased approximately 10 °C (11.24%) and then stabilized. To measure the stator, core, and heat exchanger temperatures, RTD (resistance temperature detector) sensor arrays were used.[2] The stator and core temperatures stabilized after 3 and 2.5 hours, respectively. Observe the cold and hot air temperatures of the stator heat exchanger exhibit a profile very similar to the stator and core temperatures, respectively.

[1]A temperature whose variation is less than 0.5 °C in a period of less than one hour was considered as stabilized.

[2]In Fig 4.24, we show the average temperatures of RTD arrays.

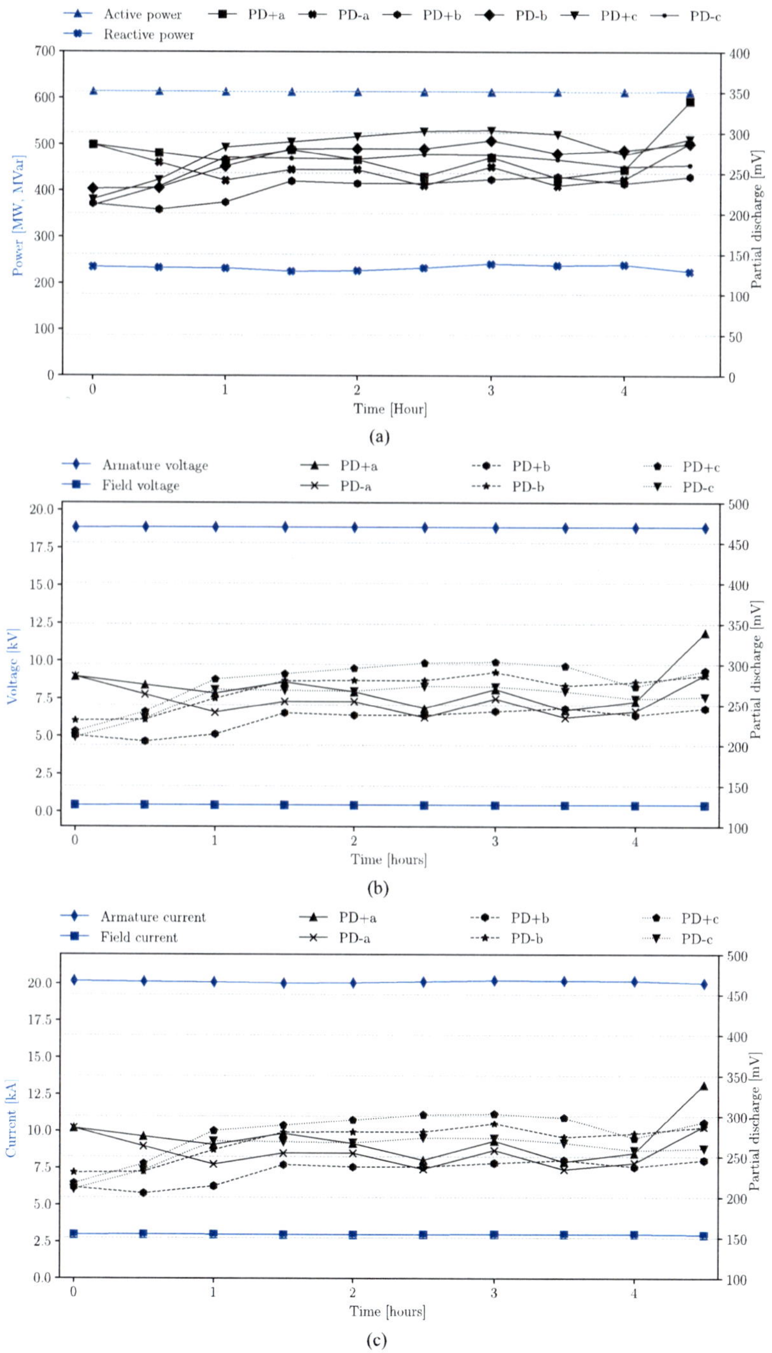

Fig. 4.23 Comparisons between PD levels (mV) and electrical variables during heating test with generator active power of 611 MW and a power factor of 0.9. (**a**) Active (MW) and reactive (MVAR) power vs. time. (**b**) Armature and field voltages vs. time. (**c**) Armature and field currents vs. time

Table 4.3 PD levels at the beginning and the end of the heating test with 100% of the nominal load. PD+: positive PD pulse; PD−: negative PD pulse

Phase	Coupler	Start of the test	End of the test	Increase
A	PD+a (mV)	285.31	339.22	19%
	PD-a (mV)	285.31	287.03	1%
B	PD+b (mV)	212.03	245.63	16%
	PD-b (mV)	230.47	286.56	24%
C	PD+c (mV)	217.26	292.50	35%
	PD-c (mV)	210.00	260.00	24%

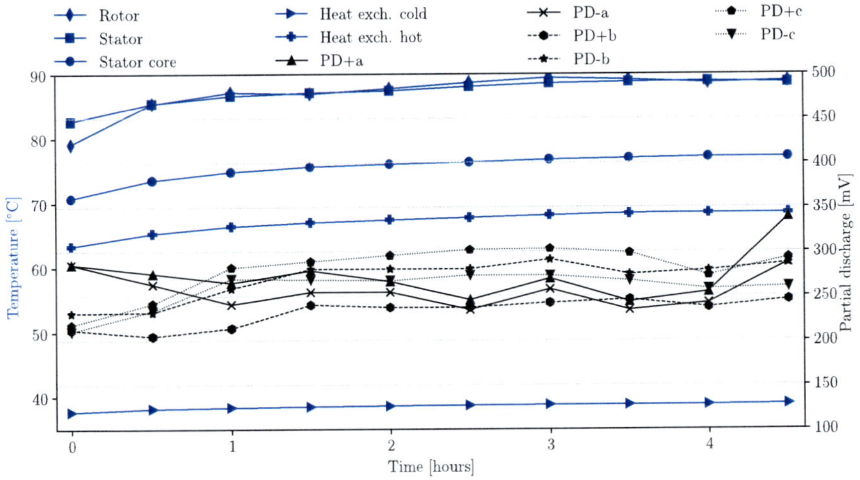

Fig. 4.24 Comparisons between PD levels (mV) and temperatures during heating test with generator active power of 611 MW and a power factor of 0.9

Table 4.4 Pulses counting at the beginning and at the end of the heating test

Phase	#Pulses (beginning)	#Pulses (end)	Increase
A	415	1097	264%
B	522	1310	251%
C	561	1021	182%

Phase-Resolved PD Analysis

Even though the amplitude levels of PDs remained almost constant—between 200 and 300 mV—the number of PD pulses increased as the heating test proceeded. Table 4.4 presents the number of PD pulses detected at the beginning and at the end of the heating test. Figure 4.25 allows us to compare the PRPDs for each phase before and after the heating test. Note that the increase in PD counting does not imply a change in the main characteristic of the PRPD pattern, in this case, typical of internal voids.

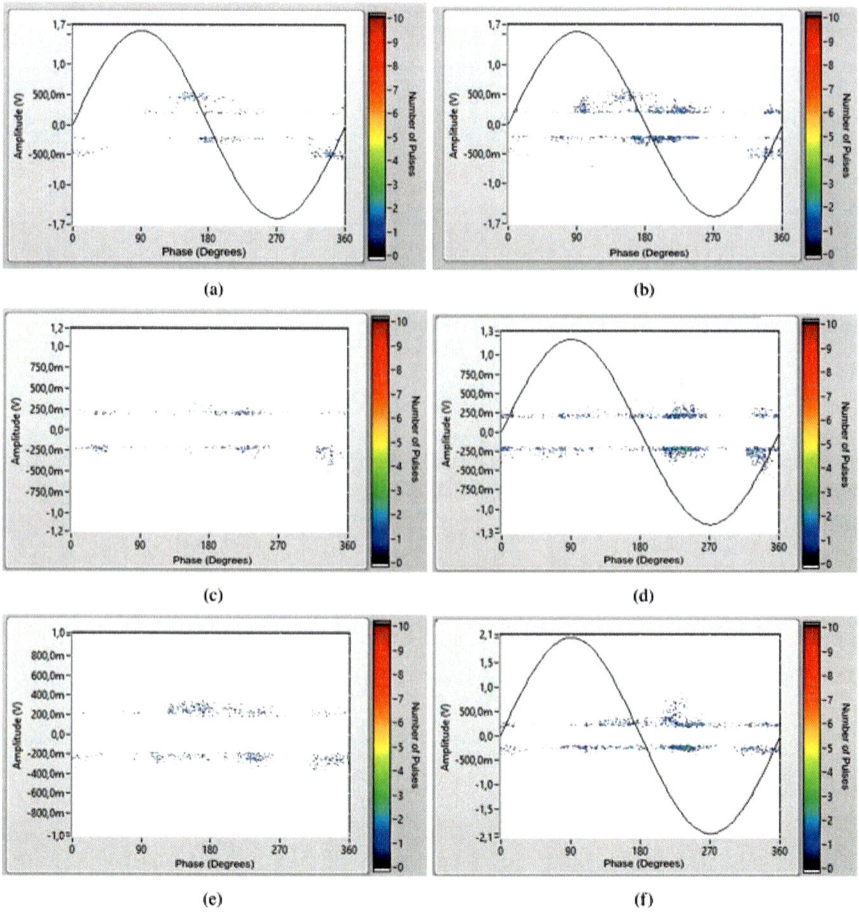

Fig. 4.25 PRPD patterns from phases A, B and C at the beginning and at the end of the heating test. (**a**) PRPD of phase A at the beginning of the test. (**b**) PRPD of phase A at the end of the test. (**c**) PRPD of phase B at the beginning of the test. (**d**) PRPD of phase B at the end of the test. (**e**) PRPD of phase C at the beginning of the test. (**f**) PRPD of phase C at the end of the test

4.4.4 PDs Evolution Over Time

Figure 4.26 shows the generator measurement history of PDs since its first excitation until after 2 months of commercial operation. Active power during operating range and heating tests, and the corresponding PD levels for phase A are depicted. Note the rightmost measurements for Power, PD-A, and PD+A were taken after 2 months of commercial operation of the generator (the heating test finishes at the measurements before the last ones, as indicated by the corresponding bar in Fig. 4.26). In that final verification, PD magnitudes are slightly lower with regard to the ones registered during the heating test. This happens because in the heating test the unit was

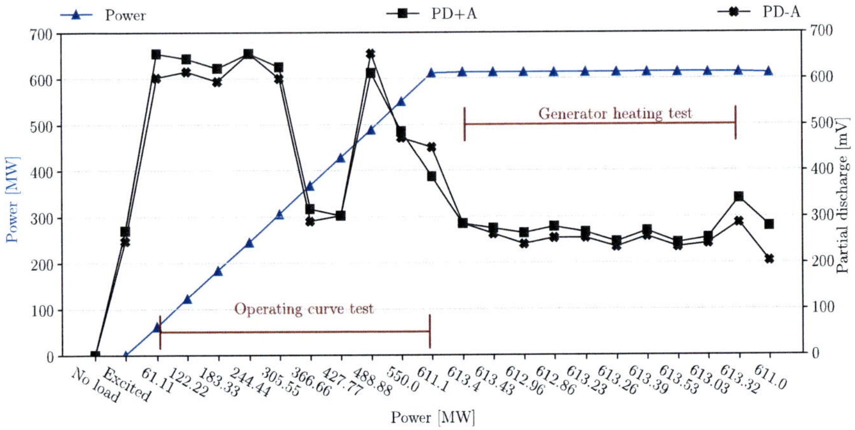

Fig. 4.26 PDs and generator load data obtained during commissioning tests (operating curve and heating tests) and after 2 months of commercial operation (leftmost measurement)

operating overexcited, with a power factor of 0.9. The significant variation of PD levels observed throughout the commissioning tests reinforces the need to consider a large set of factors, such as vibration, temperature, load cycle, among others, when monitoring PD [20].

4.5 Final Remarks

This chapter presented results and procedures regarding PD measurements in the field, performed in hydroelectric plants. We verified how the availability of a historical log of PD data favors machine diagnosis by providing the means to properly evaluate the severity of PD levels. In this sense, best practices in predictive maintenance consider online monitoring, where PD measurements are continuously taken and analyzed. Such periodical measurements constitute the PD history of a generator unit, from which the typical PD behavior for a given stator winding can be characterized. Only then we can correctly track the evolution of PDs over time.

We also discussed PD monitoring made during the commissioning tests of a new generator, comprising its first excitation, synchronization, operating range and heating tests, and a late PD measurement after 2 months of commercial operation. Our experimental results confirmed the correlation between PD activity and other factors such as vibration, temperature and load cycle [1], and the importance to consider them in rigorous PD monitoring of hydrogenerators.

References

1. Rotating Electrical Machines - Part 27-2: On-Line Partial Discharge Measurements on the Stator Winding Insulation of Rotating Electrical Machines. IEC/TS 60034-27-2. International Electrotechnical Commission, Tech. Rep. (2012).
2. Gross, D. W. (2003, June). Partial discharge diagnosis of motor defects. In *Nordic Insulation Symposium*, Tampere.
3. Stone, G., Sasic, M., Dunn, D., & Culbert, I. (2009). Recent problems experienced with motor and generator windings. In *2009 Record of Conference Papers - Industry Applications Society 56th Annual Petroleum and Chemical Industry Conference* (pp. 1–9).
4. International Electrotechnical Commission. (2015). Rotating electrical machines - part 27-3: Dielectric dissipation factor measurement on stator winding insulation of rotating electrical machines. International Electrotech- nical Commission, Geneva, Tech. Rep. Edition 1.0.
5. Stone, G., & Warren, V. (2005). Differences in stator winding partial discharge activity between manufacturers. In *Proceedings of the XIVth International Symposium on High Voltage Engineering*.
6. Iris Power, *Iris Power website*. Accessed Jan 24, 2023. http://www.irispower.com
7. Power Diagnostix Systems. (2023). *Power Diagnostix Systems website*. Accessed 24 Jan, 2023. http://www.pd-systems.com
8. Belec, M., Hudon, C., & Nguyen, D. (2006). Statistical analysis of partial discharge data. In *Conference Record of the 2006 IEEE International Symposium on Electrical Insulation* (pp. 122–125).
9. Stone, G. (2005). Partial discharge diagnostics and electrical equipment insulation condition assessment. *IEEE Transactions on Dielectrics and Electrical Insulation, 12*(5), 891–904.
10. Bartnikas, R., & McMahon, E. (1979). *Engineering Dielectrics: Corona Measurement and Interpretation - STP 669*. ASTM special technical publication (Vol. 1). West Conshohocken: American Society for Testing and Materials.
11. Sedding, H., Campbell, S., Stone, G., & Klempner, G. (1991). A new sensor for detecting partial discharges in operating turbine generators. *IEEE Transactions on Energy Conversion, 6*(4), 700–706.
12. Kelen, A. (1976). *The functional testing of hv generator stator insulation*. Accessed Jan 23, 2023. https://e-cigre.org/publication/15-03_1976-the-functional-testing-of-hv-generator-stator-insulation
13. IEEE guide to the measurement of partial discharges in rotating machinery (2000, August). *IEEE Std 1434-2000* (pp. 1–64).
14. Fluke Corporation. (2002). *ScopeMeter 190C/190B Series and ScopeMeter 120 Series: Technical data*. Accessed Jan 24, 2023. https://dam-assets.fluke.com/s3fs-public/1629083_.pdf.
15. Bartnikas, R., & Levi, J. H. E. (1969). A simple pulse-height analyzer for partial-discharge-rate measurements. *IEEE Transactions on Instrumentation and Measurement, 18*(4), 341–345.
16. Carvalho, A. T., Amorim Junior, H. P., Rodrigues, C. F. C., Brasil, F. S., Vilhena, P. R. M., & Carvalho, D. D. (2017). Virtual instrumentation for partial discharge monitoring. In *Electrical Insulation Conference An IEEE DEIS Sponsored Conference, Baltimore. Partial Discharge* (pp. 173–176).
17. Stone, G., Culbert, I., Boulter, E., & Dhirani, H. (2014). *Electrical insulation for rotating machines: Design, evaluation, aging, testing, and repair* (2nd ed.). Hoboken: Wiley-IEEE Press.
18. Aakre, T., & Ildstad, E. (2020). PD-activity in generator stator bar insulation versus voltage frequency and temperature. In *2020 IEEE 3rd International Conference on Dielectrics (ICD)* (pp. 870–873).

19. IEEE guide for test procedures for synchronous machines part I — acceptance and performance testing part II — test procedures and parameter determination for dynamic analysis - redline. *IEEE Std 115-2009 (Revision of IEEE Std 115-1995) - Redline* (pp. 1–219, 2010).
20. Stone, G. C. (2012). A perspective on online partial discharge monitoring for assessment of the condition of rotating machine stator winding insulation. *IEEE Electrical Insulation Magazine, 28*(5), 8–13.

Chapter 5
Numerical Modelling and Pinpointing of Partial Discharges

5.1 Introduction

In order to maintain the functionality of high voltage equipment, it is crucial to ensure the quality of the insulation [1–11]. High-level electric fields within this apparatus cause the insulating device parts to deteriorate. In order to partial discharges (PDs) develop inside insulation, degradation must favor the formation of dielectric heterogeneities. The dielectric insulation of the generator may get weakened by the many discharges that might happen during each cycle of operation. It has been established via research on such discharges (see papers by Wang et al. [8], Wenzel et al. [11] and Birlasekaran [7] that the transient fields they create may be utilized to determine the isolation state.

Different mathematical methods have been proposed during the past 20 years for localizing discharges in high voltage equipment [3–8, 10–12]. The paper [8] used neural network techniques to identify patterns relative to PDs. In these experiments, bars were constructed specifically to observe different types of discharges: lamination, winding and slotting. In [11], the authors applied fuzzy logic to analyze PD transients in transformers and wavelet transforms were used in [7] to pinpoint discharges.

In [6], the authors presented a technique called *Intech* which allows for more precise identification of the region of the generator producing the discharges. In [3], the authors applied the particle swarm optimization (PSO) method for diagnosing power transformers. Several measurement techniques have also been presented, such as in [4], where optical sensors were used to perform localization of single and multiple discharges in high voltage equipment. Other papers, such as [4, 10], use acoustic measurements combined with different electrical methods to localize discharges in transformers.

In this chapter, we review a method developed by the book authors for localizing numerous PDs utilizing the transient voltage spectrum recorded at a single location on a hydrogenerator coil. The fields associated to the voltages have their source

V. Dmitriev et al., *Partial Discharges in Hydroelectric Generators*, Power Systems,
https://doi.org/10.1007/978-3-031-36604-8_5

at the PDs excitation points. The method entails mapping frequencies related to the maximum and minimum of the signal spectrum obtained by a sensor and connecting them to the locations where discharges occur. This strategy is supported by the natural resonances that the examined structure [13, 14] produces, which are also dependent on the locations of the electromagnetic sources [15]. Experimentally and statistically, the resonances are studied.

The fundamental benefit of this technology is the ability to use the superposition principle to recognize many simultaneous discharges, including those happening in different coils. Numerical analysis supports the suggested method's efficacy. Precision of the author's FDTD-3D simulator [16], which is based on finite-difference time-domain method [17, 18], is verified by numerically replicating existing experimental setups from the literature. Details on FDTD simulation method and its computer implementation are given in Appendix C.

This chapter also conducts experimental research. By contrasting experimental and numerical data, a controlled local PD injection schema is put forth and proven to work.

5.2 Rectangular Bar Electromagnetic Wave Propagation Resonance Analysis

A rectangular bar was modeled (Fig. 5.1) with the intention of studying the influence of bar parameters on propagation of electromagnetic signals produced by partial discharges taking place at various points in the insulation layer in order to verify the fundamental physical aspects of the problem. The bar is 3 meters long (Fig. 5.1a), and it is made up of three layers: an inner metal bar (copper), a dielectric layer (mica), and an outside layer that is semi-conductive (Fig. 5.1b).

Excitation of the Ez component was used to simulate separate instances of z-polarized partial discharges at three points (P1, P2, and P3) of the FDTD-3D model of the bar (Fig. 5.1), and the induced transient voltage was captured by a sensor at the location shown in Fig. 5.1a. By adjusting the electrical permittivity, conductivity, and rising time of the excitation signal, several simulations were performed.

The magnetic field (rather than electric field) propagation caused by partial discharges simulated at P1, P2, and P3 for three separate instants is shown in Fig. 5.2 for the best field visualization. ε_r is equal to 5.4, $\sigma = 0.00010394$ S/m, and $\mu_r = 1$ [14, 19].

In Fig. 5.2, it is easy to see how the PD location affects the different wavelengths that are generated. This occurs as a result of natural electromagnetic resonances being produced by field reflections at the bar's ends and the distances involved in this electrodynamics process (from the PD location to the bar ends). Due to field reflections at the metallic sections of the bar, additional resonances appear.

The rising time τ of the discharge pulse is the first parameter examined. It was possible to replicate pulses with rise times of 0.5, 1, 3, and 5 ns [9]. The parameters

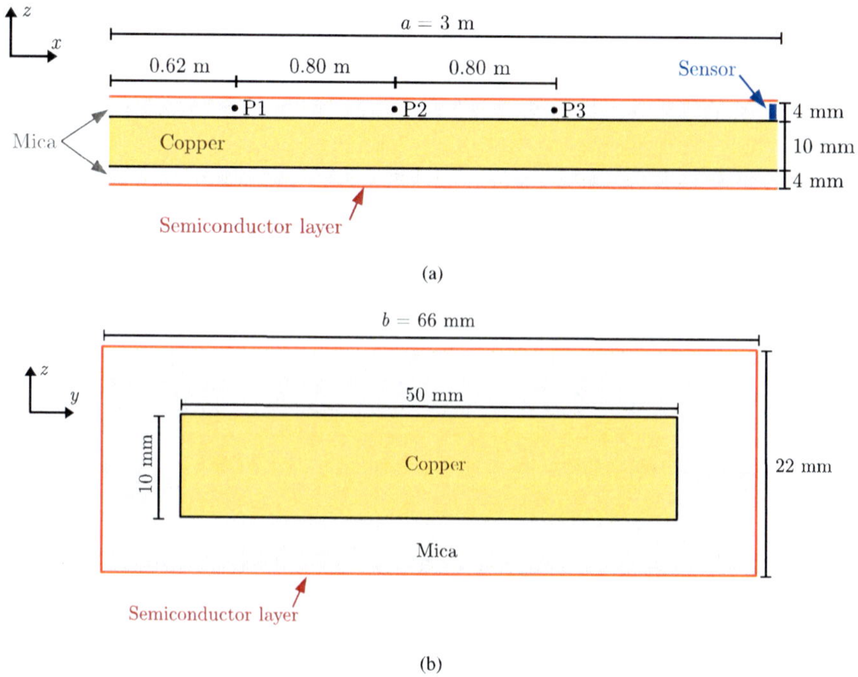

Fig. 5.1 Sections of the rectangular bar: (**a**) longitudinal and (**b**) transversal

$\varepsilon_r = 5.4$, $\sigma = 0.00010394$ S/m, and $\mu_r = 1$ were used to simulate the mica insulator for this analysis [14, 19].

Spectral maxima and minima were discovered to be fundamentally dependent on the structure itself and, of course, the PD's location since they do not change when tau is altered, as one can see in Fig. 5.3a$_1$, b$_1$ and c$_1$. This is connected to the field reflections at the ends of the bars that were previously mentioned.

The field distributions of Fig. 5.3a$_2$–a$_5$, b$_2$–b$_5$ and c$_2$–c$_5$ show different wavelengths for this instance. It should be noted that the signal intensity received by the sensor conveys energy in the high frequency band of the examined spectrum for rising times of 0.5 ns and 1.0 ns. For frequencies starting about 600 MHz, the signal amplitudes are larger as a result. As long as the PD pulse has energy in the studied spectral range , you will see that the frequency points at which the peaks and minima occur in the spectrum are kept.

Additionally, numerical research was done to examine the impact of the dielectric's electrical conductivity, σ. Three simulations were run for each of the three excitation points (P1, P2, and P3) using various σ values [19]: 5.2×10^{-5}, 1.04×10^{-4} and 2.07×10^{-4} S/m, with $\varepsilon_r = 5.4$ and $\mu_r = 1$.

It has been shown that changing σ does not affect the frequencies at which peaks and minima in the spectrum occur. The changes in spectra that are seen as a function of σ are exclusively related to spectral amplitudes , i.e., when σ increases,

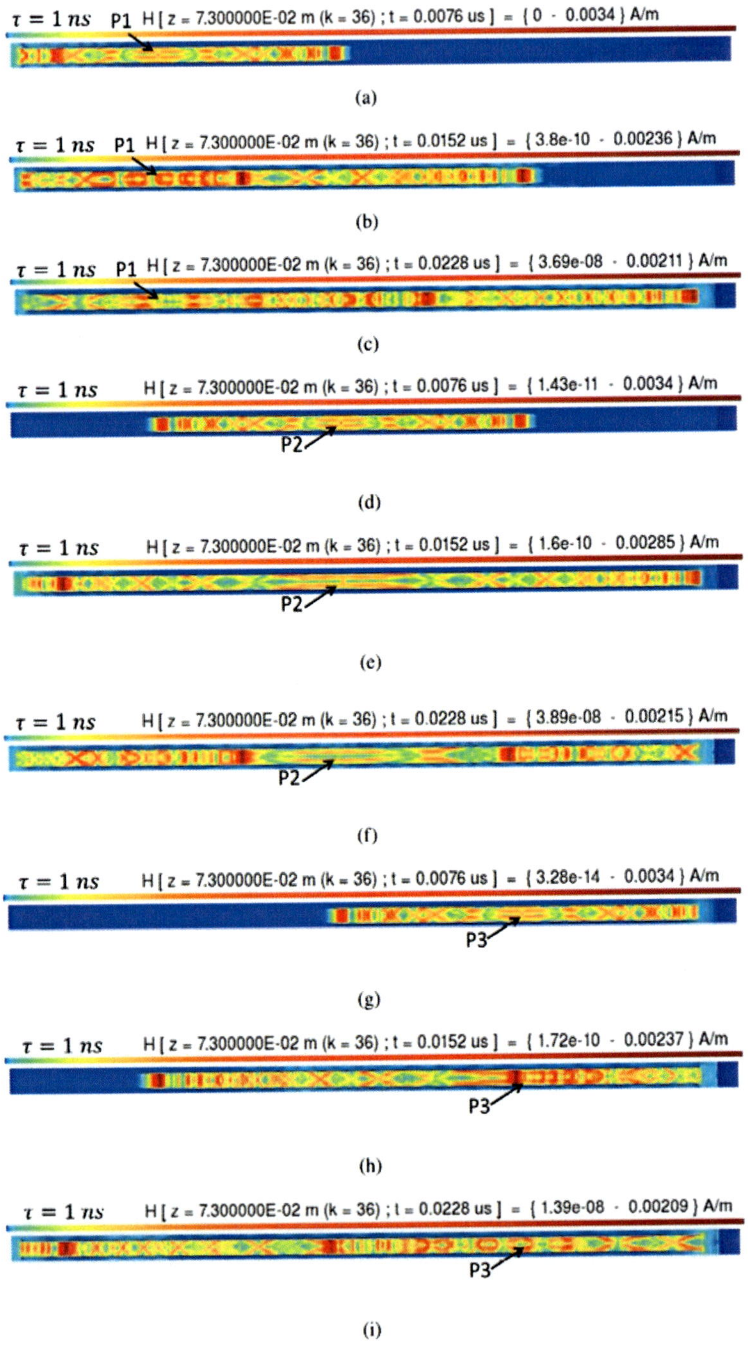

Fig. 5.2 Visualization of magnetic field propagation for PD taking place at (**a**) P1 for $t = 0.0076\,\mu\text{s}$, (**b**) P1 for $t = 0.0152\,\mu\text{s}$, (**c**) P1 for $t = 0.0228\,\mu\text{s}$, (**d**) P2 for $t = 0.0076\,\mu\text{s}$, (**e**) P2 for $t = 0.0152\,\mu\text{s}$, (**f**) P2 for $t = 0.0228\,\mu\text{s}$, (**g**) P3 for $t = 0.0076\,\mu\text{s}$, (**h**) P3 for $t = 0.0152\,\mu\text{s}$ and (**i**) P3 for $t = 0.0228\,\mu\text{s}$. The dielectric parameters are $\varepsilon_r = 5.4$, $\sigma = 0.00010394$ S/m and $\mu_r = 1$

Fig. 5.3 Comparison of the spectra of the received signal for different rise time: (**a**) for partial discharge placed at P1, (**b**) at P2 and (**c**) at P3

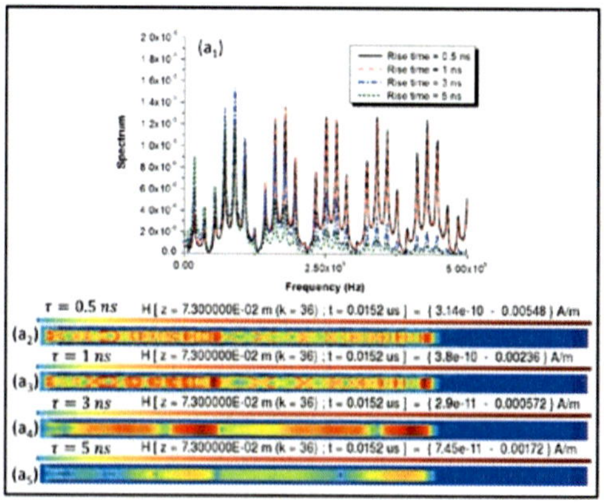

(a)

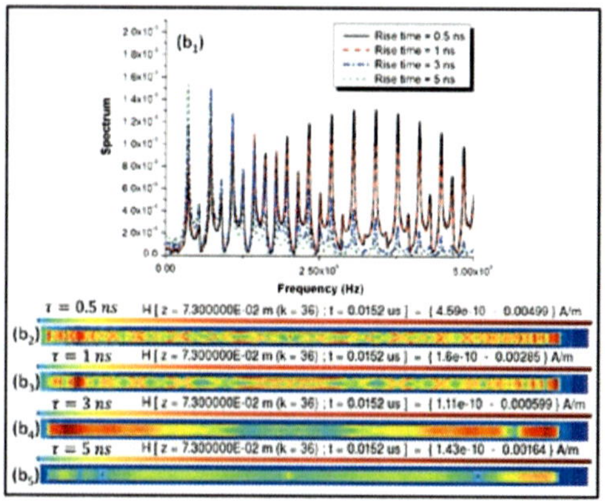

(b)

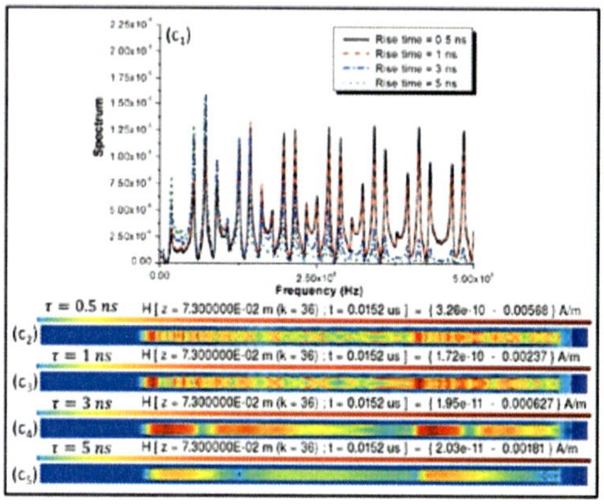

(c)

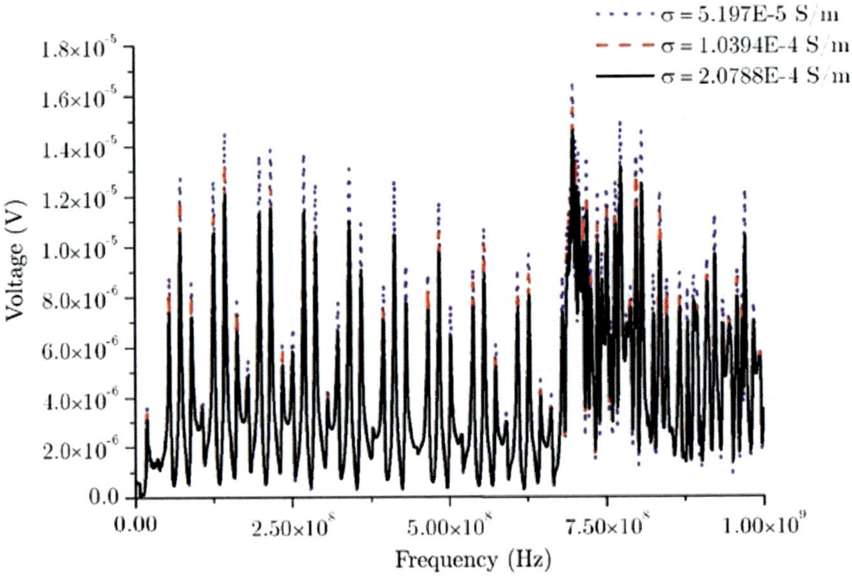

Fig. 5.4 Comparisons of spectra of received signals for different values of σ for PD excitation at P3. Similar behaviour is seem when excitation is placed at P1 or P2

currents tend to be surface-level and the interior electric field is weakened. The comparison of the received signal spectra for excitations at P1, P2, and P3 in Fig. 5.4 demonstrates the effect that was just referred to.

The spectrum effects of mica insulation relative electrical permittivity were then assessed. The partial discharge simulation seen in Fig. 5.1a was repeated with the dielectric's relative permittivity being varied between 5.4 and 8.0 [14].

Resonance frequencies f_r in the spectral range between 80 and 400 MHz were analytically confirmed using the bar in Fig. 5.1 as a waveguide. The resonance frequencies of the bar were determined analytically using Eq. (5.1) [13] for the aim of validating numerical solutions. In Fig. 5.5, the frequencies calculated by Fig. 5.5 and via FDTD simulation are compared. Resonance frequencies may be calculated using the equation [13]

$$f_r = \frac{u'}{2}\sqrt{\left[\frac{m}{a}\right]^2 + \left[\frac{n}{b}\right]^2 + \left[\frac{p}{c}\right]^2}. \tag{5.1}$$

In (5.1), $u' = C_0/\sqrt{\varepsilon_r}$ is the electromagnetic wave velocity in mica. Additionally, $C_0 = 299,792,458$ m/s is the speed of light in free space and $\varepsilon_r = 7$ is the relative permittivity of mica. In Fig. 5.1a, b and c are the waveguide's dimensions, while m, n and p are integers. Concordance between the FDTD results and (5.1) is seen in Fig. 5.5 for resonance frequencies.

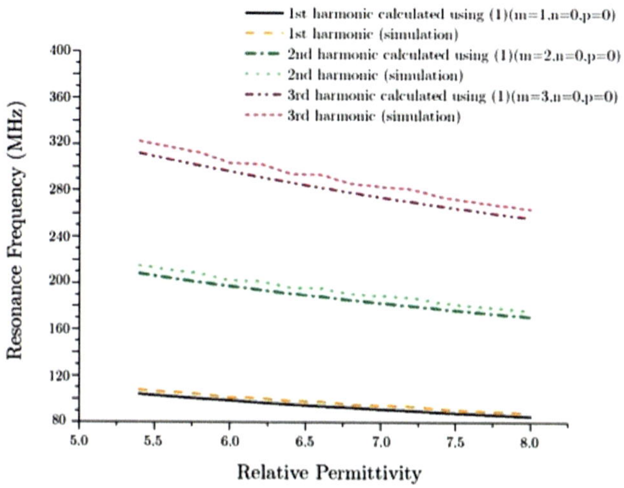

Fig. 5.5 Comparison of resonance frequencies calculated using Eq. (5.1) and frequencies obtained via FDTD simulation

5.3 Description of the Problem and Numerical Modeling

Design schematics from Eletronorte for a synchronous generator winding with 48 prominent poles has been used as basis for conceiving the numerical models. The machine has the following nominal specs: 30,402 kVA of power, 13,200 V of operational voltage, 5% operational range, 1330 A of current oscillating at 60 Hz and 0.95 power factor. The project files for the equipment, [15] (Fig. 5.6a), were used to create a comprehensive FDTD model of a stator coil, which is shown in Fig. 5.6b. To transfer this geometric data to the SAGS FDTD simulator, implemented by authors [16] and modified to simulate PDs, a particular computational procedure was created. The Appendix contains the results of the numerical validation of this program for the specified problem, which was carried out by comparing the results with data from the literature.

As shown in Fig. 5.6, the FDTD numerical model of the bar is composed of three layers: an inner metal structure, a dielectric layer surrounding it, and an outer metallic layer. The (average) parameters $\varepsilon_r = 7$, $\sigma = 10^{-15}$ S/m, and $\mu_r = 1$ define the dielectric layer (mica) [13]. Copper is modelled as a metallic element with parameters $\varepsilon_r = 1$, $\sigma = 5.8 \times 10^7$ S/m and $\mu_r = 1$), as defined in [14]. We have used $848 \times 254 \times 165$ cubic Yee cells to represent the entire volumetric space considered. Spatial step Δ was set to 2 mm.

The waveform and spectrum of the voltage source used to mimic the discharges are shown in Fig. 5.7a and b, respectively. The greatest edge permitted for Yee's cell for lowering numerical dispersion to acceptable levels is $\Delta_{max} = \frac{1}{10}(C_0/\sqrt{7})/(1 \times 10^9) \approx 11.3$ mm as long as the pulse's highest significant frequency is around 1 GHz (Fig. 5.7b). Notice that $\varepsilon_r = 7$ in mica is the highest relative permittivity in the

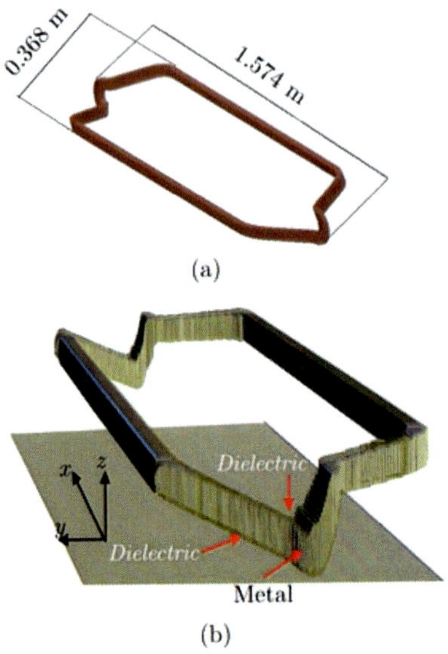

Fig. 5.6 (**a**) 3D view of hidrogenerator machine coil modelled in SolidWorks®; (**b**) inner structure of the model represented by the SAGS FDTD simulator

physical domain. We can see that the employed numerical grid is appropriate for representing the electromagnetic waves created by the pulse of Fig. 5.7 , since the edges of each Yee cell in the grid were set to 2 mm (smaller than one tenth of the minimum propagating wavelength [18]).

5.4 A Spectral Method for Examining Stator Coils in a Laboratory Environment

As previously mentioned, a particular coil may simultaneously experience many partial discharges at various points. As a result, cases where modelled in which pairs of simultaneous discharges occur in the quadrants depicted Fig. 5.8. We also took into account discharges created by a single quadrant. Partial discharge signals spread throughout the structure and experience several reflections. Due to the relative permittivity ε_r of the insulation, the signals' propagation velocities are decreased when compared to the speed of light in a vacuum by a factor of $1/\sqrt{\varepsilon_r}$. Additionally, waves naturally expand, which attenuates signals, as also does mica's poor electrical conductivity. Thus, the location of the discharge(s) has a significant impact on the transient signals that the sensors detect (see Fig. 5.8).

It is confirmed that every portion of the coil produces a distinct spectral pattern, in which local maxima and minima characterize a given spatial region due to produced specific resonances, taking into account that each pair of transmitter (PD) and receiver (sensor S) are associated to a unique propagation channel (this physical aspect is verified experimentally in Appendix, Section C). A computer method

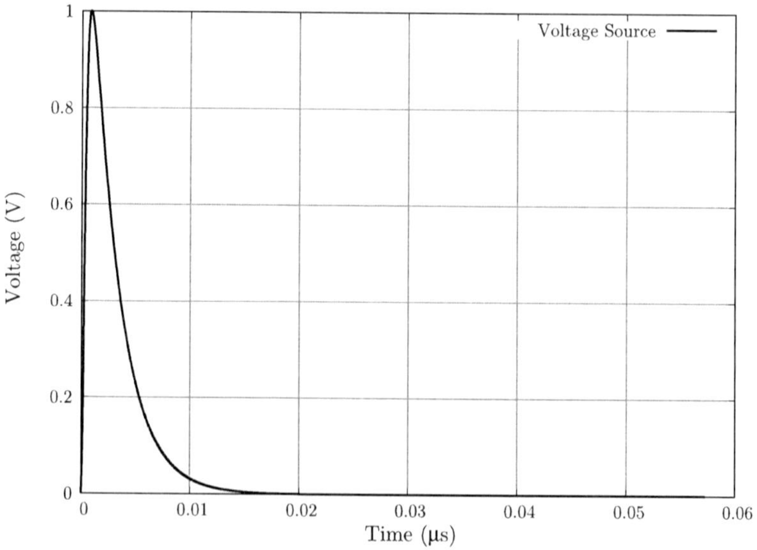

(a)

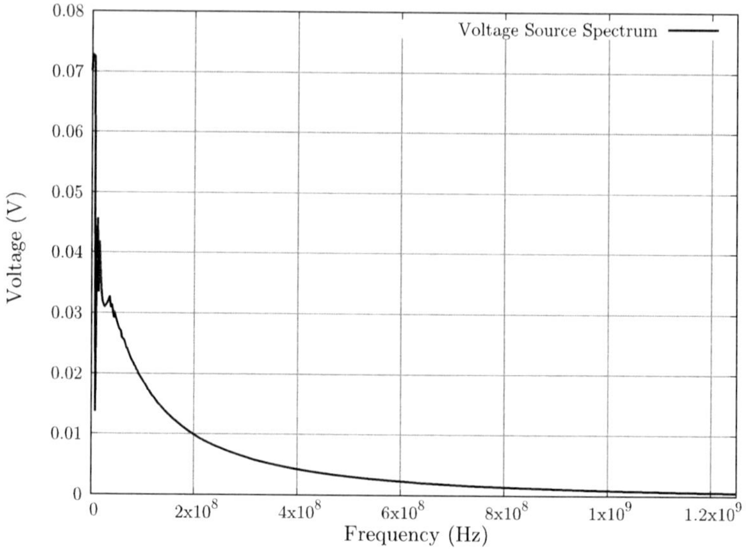

(b)

Fig. 5.7 (**a**) Normalized time domain waveform used as FDTD PD excitation source, (**b**) excitation source spectrum

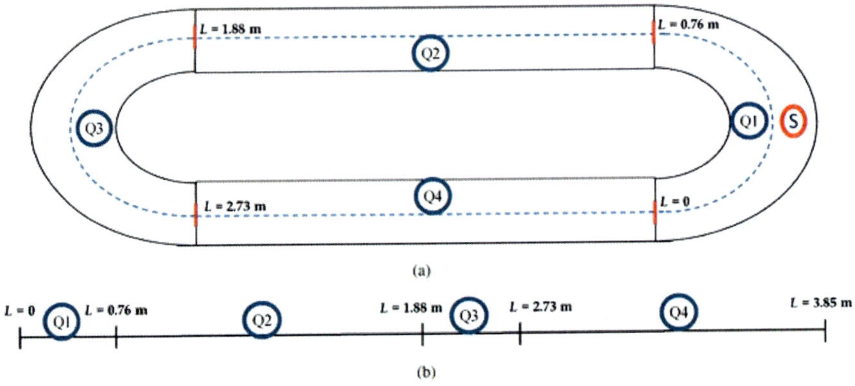

Fig. 5.8 (**a**) Cross section of bar and relation of L with positions in structure regions; (**b**) L variable in a one-dimensional coordinates system

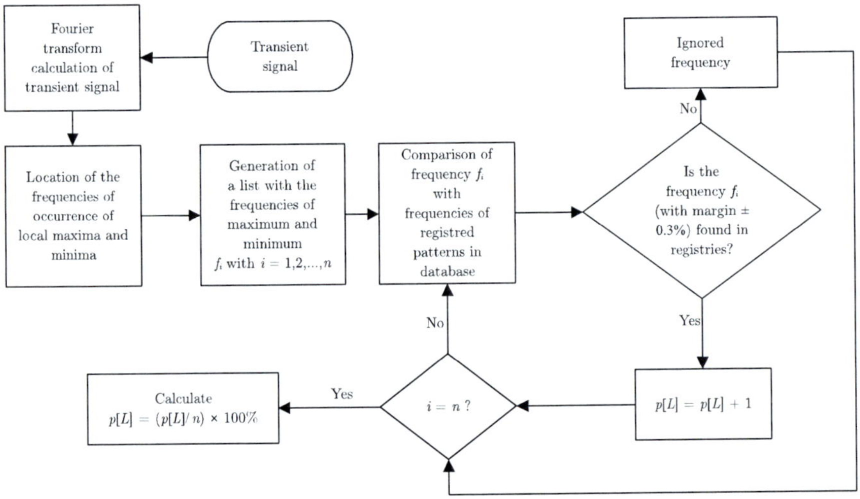

Fig. 5.9 Algorithm for proposed PD pinpointing technique for a fixed value of L

(Fig. 5.9) was designed and put into use using this information to determine the frequencies at which local peaks and minima in the spectral functions occur . Such resonance frequencies and the coil coordinates L, linked to transmission spots and a particular receiver, were compiled into a database. The location(s) from which discharges occur are inferred using the database (Fig. 5.5).

Following the establishment of the database, two simultaneous discharges were simulated and placed in various locations. The Fourier transforms of the signals collected by the sensor were then computed (Fig. 5.9). The resonance frequencies of the transients that were received were searched for in the database.

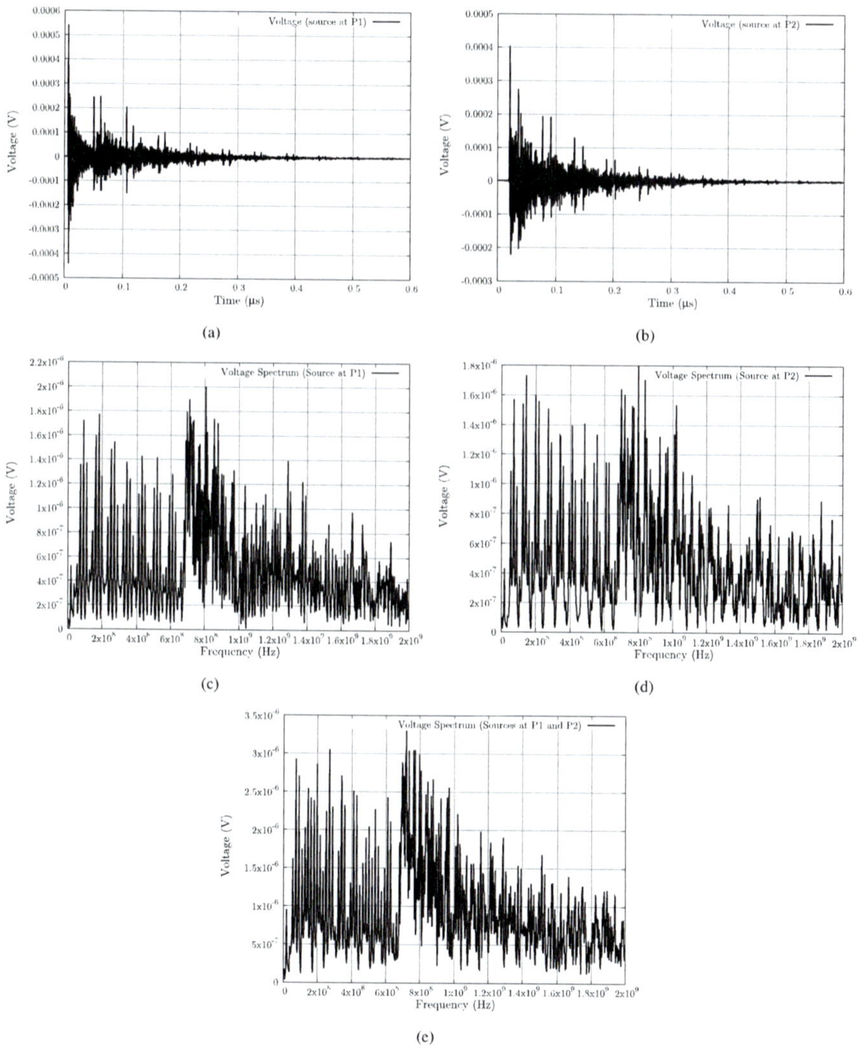

Fig. 5.10 Voltage signals of simulation 161 registered by the sensor S. For discharge 1, we have: (**a**) the registered transient signal and (**b**) its frequency spectrum. For discharge 2, we have: (**c**) the registered transient signal and (**d**) its frequency spectrum. The simultaneous occurrences of discharges 1 and 2 produce the spectrum given in (**e**)

When numerous discharges at various sites occur simultaneously, their distinct spectra are physically summed together due to the superposition principle in Maxwell's equations and the Fourier transform (see Fig. 5.10). With the help of this feature, it is possible to look for clusters of resonances in the spectrum of concurrent PDs that represent the distinct contributions of each PD happening at specific point.

In fact, the superposition of the numerous discharge spectra can alter (or perhaps completely cancel out) some resonance information. However, this may be handled by specifying the likelihood that a PD will occur at a certain location on the structure. The total number of resonances $C[L]$ listed in the database for a PD flashing at location L that are also present in the observed spectrum may be used to determine this probability, $p[L]$ (consequence of an arbitrary number of occurring PDs). In this regard, we may write the mathematical expression

$$p[L] = \frac{C[L]}{n[L]} \times 100, \tag{5.2}$$

in which $n[L]$ is the total amount of resonances for a discharge occurring at L that are recorded in the database. The diagram in Fig. 5.9 provides an algorithmic explanation of this concept. A map of the statistical probabilities for discharge localization is then made by running the method again for different values of L.

An integer number is used as index for every modeled pair of partial discharges. For instance, the points stimulated by partial discharges for simulations with indices 5 and 161 are shown in Fig. 5.11b.

Figure 5.10a and b, for the simulation 161, show the transient signals captured by the S sensor for discharges occurring in the dielectric material. Figure 5.10c and d illustrate their spectra, respectively. According to the superposition principle, when two pulses occur at the same time, their spectra and temporal signals are linearly combined (Fig. 5.10e). Comparing Fig. 5.10c–e, it can be seen that the majority of the local minima and maxima seen in Fig. 5.10c and d are preserved in the graph of Fig. 5.10e, allowing the identification of the various discharges. Since solely the Fourier transform's magnitude is taken into account (phase is being disregarded), it is also notable that this is true even when the discharges do not start at the exact same time.

5.5 Results Regarding PD Pinpointing

5.5.1 Pinpointing Multiples Discharges in Just One Coil

Initial simulations of many concurrent partial discharges for a single hydrogenerator stator coil were run. The electric field distribution on a horizontal plane, i.e. over the x-y plane, spanning the center of the coil, is depicted in Fig. 5.11a for illustration purposes. Higher field strengths are indicated by reddish shades, whilst lesser magnitudes are indicated by blue-based hues.

The obtained statistical PD pinpointing maps, resulting from the diagnostic procedure suggested in this study, are shown in Fig. 5.12. The vertical axis displays the simulation index, while the horizontal axis indicates L (m). A total of 231 numerical experiments were carried out, in which various partial discharge occurrence places along L were simulated (from zero to 3.87 m). The color map shows the values

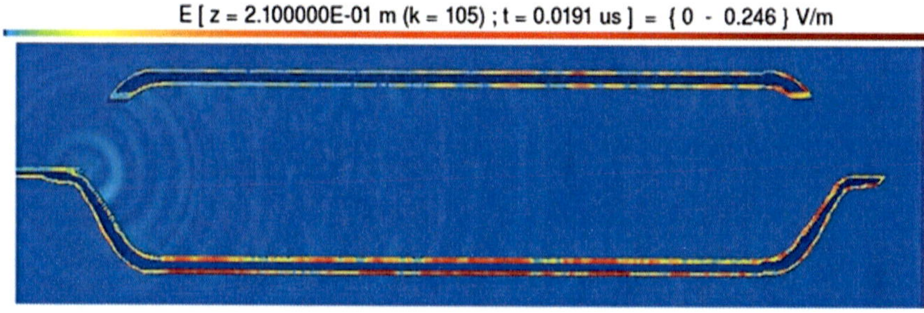

(a)

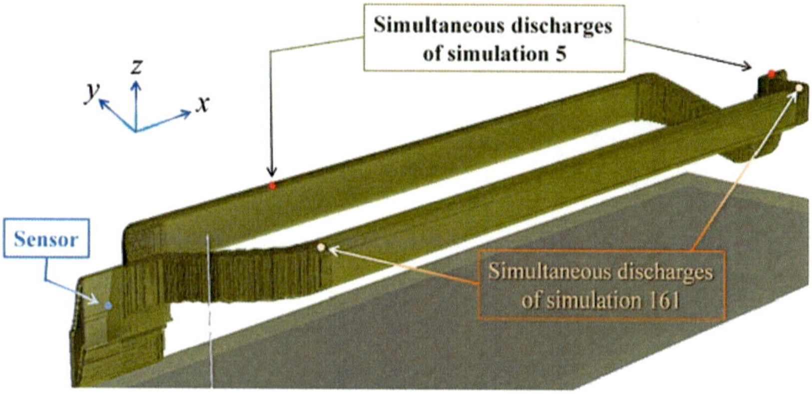

(b)

Fig. 5.11 (**a**) Visualization of electric field propagation at $t = 0.191$ μs (xy-plane that intersects the bar's centroid); (**b**) sensor location and simultaneous discharges locations of simulations 5 and 161

that the variable $p[L]$, which is a measure of the likelihood that discharges will occur, takes on. The precise locations of partial discharges as determined by the simulations are shown by the gray squares. The vertical bar displays the values that p is assuming.

In about 90% of the simulations, the algorithm proposed here generated statistical information capable of providing accurate estimates on the position of the simulated concurrent discharges. In the remaining cases, the method correctly localized the first discharge and indicated places near the real position of the second one. In 60% of cases, maximum deviations of 0.5 m from the real position of the second discharge were observed. In just 7% of the tests, imprecision between 0.5 m and 1 m was observed. In 30% of simulations, it was noted deviations from 1 m to 1.5 m. To conclude, in only 3.3% of the cases, the position could be estimated with inaccuracy greater than 1.5 m.

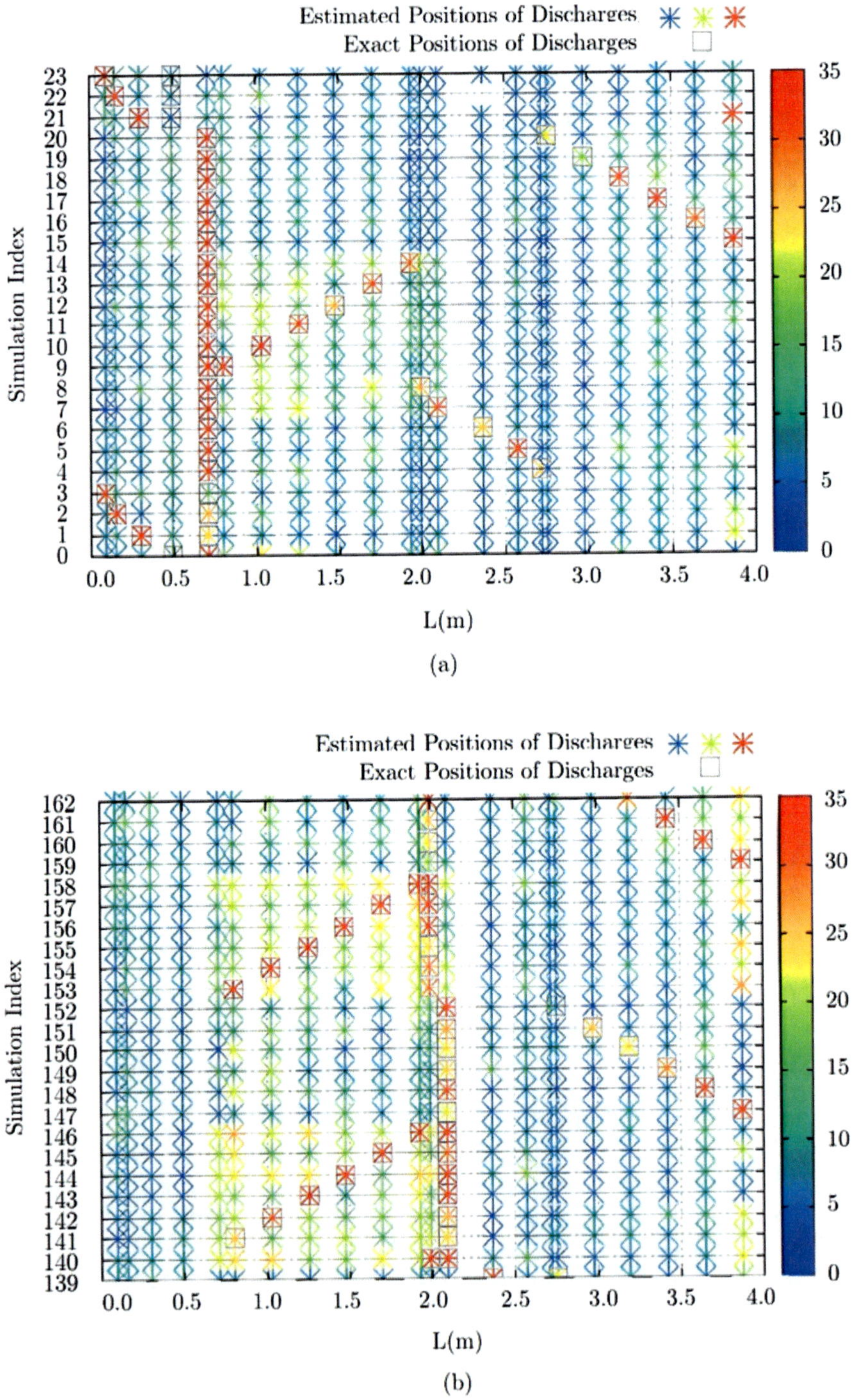

Fig. 5.12 Statistics of discharge localization for a single coil for several simulation indices: (**a**) 0–23 and (**b**) 139–162

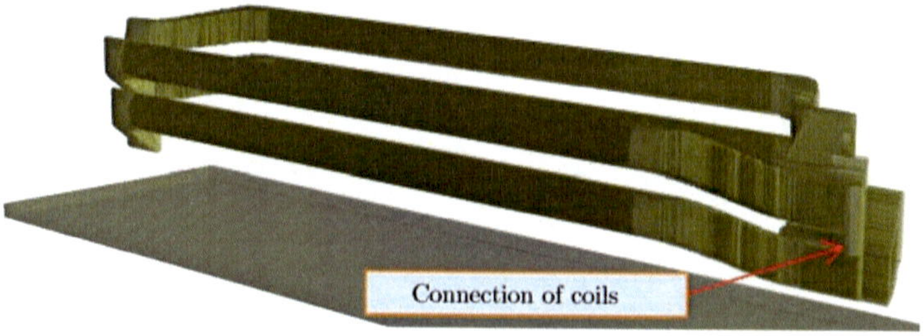

Fig. 5.13 Two adjacent connected coils

5.5.2 *Finding Multiple Discharges Occurring in Two Electrically Connected Hydrogenerator Coils*

As shown in Fig. 5.13, two identical coils from the prior example were electrically attached. For this example, a computational grid of $864 \times 254 \times 210$ cubic Yee cells with cell edges measuring $\Delta = 2$ mm was employed. The simulations' sensor placement and other features are identical to those of a single coil. It should be noted that the two coils' inner copper components were short-circuited, much like in the actual machine. Additionally, the outer metal layers and dielectric layers from the two structures were combined to form a single and continuous structure.

Obtained statistical pinpoint maps are displayed in Fig. 5.14. The results for this scenario of two neighboring coils allow for an accurate estimation of the locations of the two simultaneous discharges, even if they take place at separate coils, in roughly 81% of the simulations. In the other instances, the first discharge's location could be accurately anticipated, and the following locations were projected to be close to the second discharge's real position: In 39.7% of the tests, variations between 0.5 m and 1 m were seen; in 11.5% of the tests, deviations between 1 m and 1.5 m were detected; and in no more than 1.4% of the tests, the position of the second discharge was achieved with a deviation that was less than or equal to 0.5 m.

5.6 Laboratorial Verification of Spectral Signatures in Real Roebel Bar with Injection of Artificial PD Signals

Laboratorial tests have been carried out in this study to confirm the physical features resulting in the distinctive spectrum signatures for different propagation channels between each PD occurrence site and each sensor position, as previously discussed. We conceived a numerical model of a hydrogenerator bar that was taken from the hydroelectric power facility in Tucuruí. The bar measures 2.92 m $\times$ 18 mm $\times$ 64.3 mm and is made up of three layers: an inner cooper conductor, a mica insulator, and a semiconducting coating. Figure 5.15 shows the measurement setup.

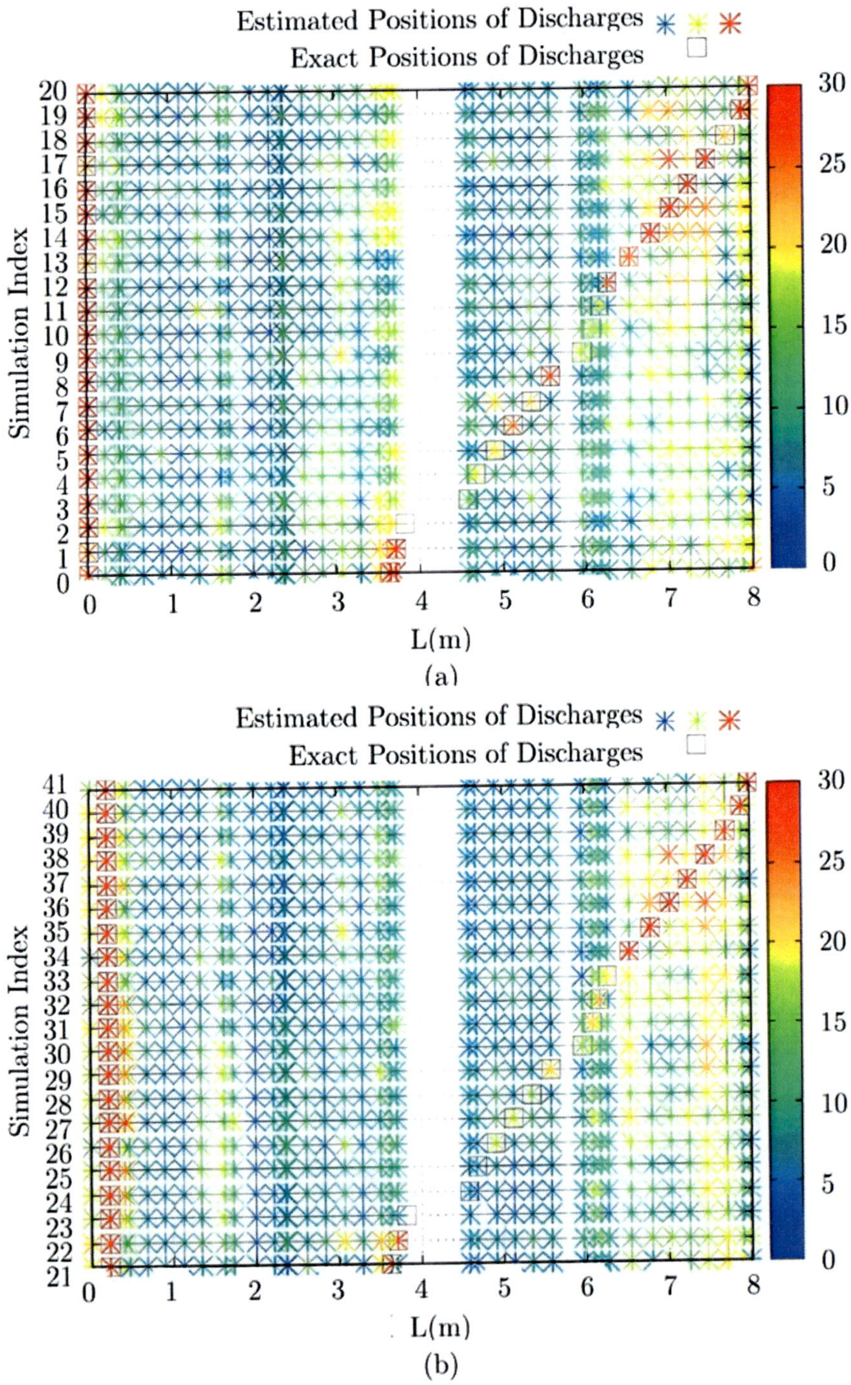

Fig. 5.14 Statistical diagnosis for two adjacent coils for several simulation indexes: (**a**) 0–20 and (**b**) 21–41

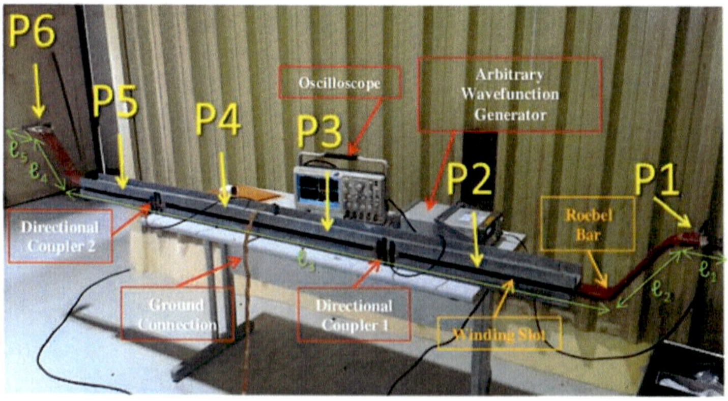

Fig. 5.15 Setup conceived for obtaining artificial PD spectral signatures

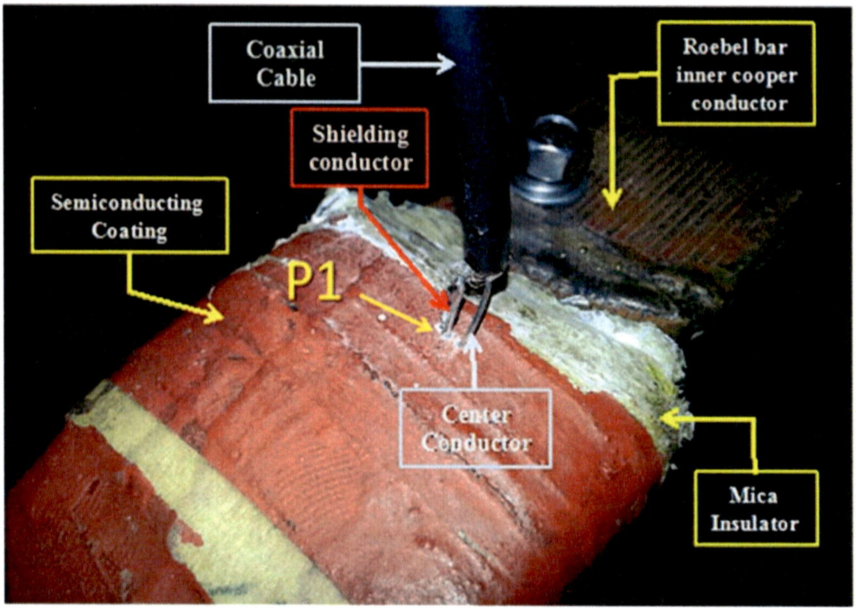

Fig. 5.16 The injection of artificial PD signal schema (illustrated at P1)

To ensure complete excitation control over the PDs sites P1–P6 (see Fig. 5.15), our arbitrary wavefunction instrument generated artificial PD signals, which were applied into the Roebel bar. The conceived experimental excitation schema is shown in Fig. 5.16.

In contrast to the experiment described in [19], the coaxial cable's center conductor penetrates the insulation in this experiment, reaching the inner cooper core of the Roebel bar, and the shielding conductor penetrates the mica insulation

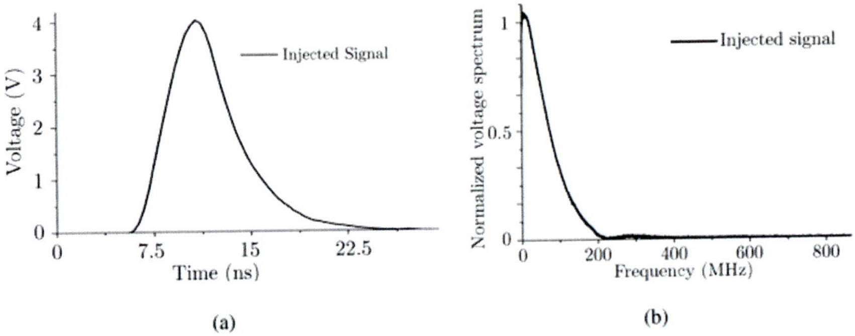

Fig. 5.17 Signal injected during experiments: (**a**) time domain waveform and (**b**) frequency domain spectrum

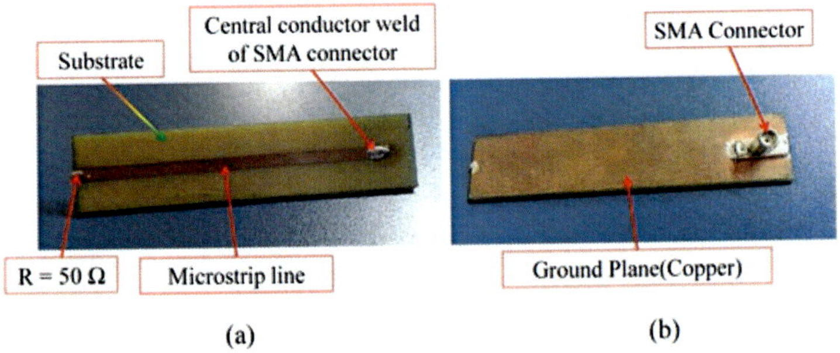

Fig. 5.18 Directional coupler: (**a**) bottom view, (**b**) top view

so that its lower end is 2 mm away from the Roebel metallic core for experimental modeling of localized PDs (see Fig. 5.16). The signal that was injected during experiments is depicted in Fig. 5.17. The distribution of signal energy is in the frequency range from a few Hertz to roughly 200 MHz, as it is shown in Fig. 5.17b.

In order to create ultrawideband field sensors, two directional couplers have been produced [20–22], which can be seen in Fig. 5.18. For suitable sensing levels, the couplers have been inserted into the structure's winding slot as indicated in Fig. 5.15. Experimental measurements and FDTD simulation calculations were used to determine the coupler return loss (S11). Observe that S11 is less than −15 dB from around 100 Hz–1.80 GHz (Fig. 5.19), and that good agreement was established between the return losses observed experimentally and numerically calculated.

The voltage spectral responses for pulse injection locations P1, P3, and P5 are shown in Fig. 5.20, which was acquired using both couplers. The observation that resonance frequencies progressively vary when the injection site is altered is crucial. The proximity (or distance) of the source to particular curves of the structure causes certain frequencies of maxima and minima to appear as long as the bar in Fig. 5.15

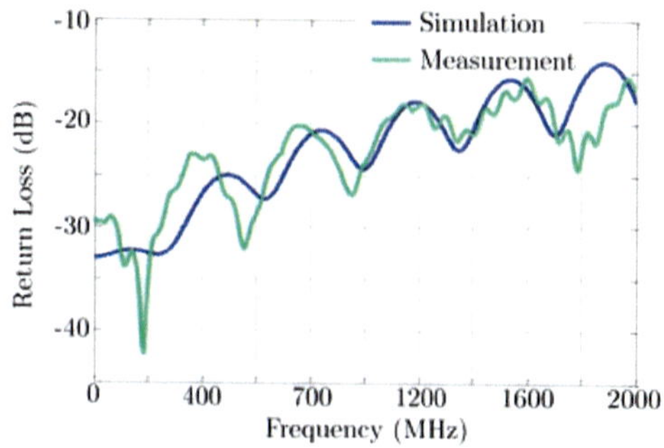

Fig. 5.19 Simulated (FDTD) and measured return losses of microstrip directional coupler

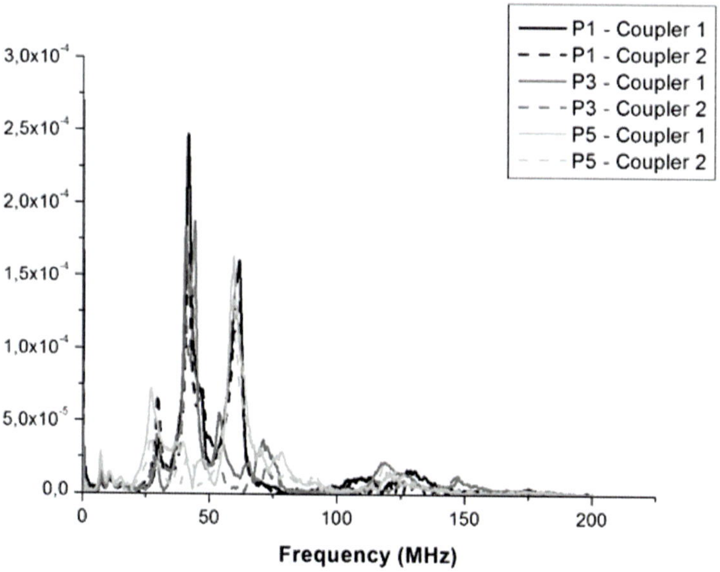

Fig. 5.20 Spectra of experimental voltage signals received by directional couplers 1 and 2 for artificial PD injection point at P1, P3, and P5

is not prismatic or symmetrical (it is made up of numerous quasi-prismatic parts). As seen in Fig. 5.15, a quasi-prismatic section of the bar (ℓ_1–ℓ_5) produces distinct resonances because of its specific dimensions.

We numerically recreate the experiment shown in Fig. 5.21 in order to test the local PD excitation schema established in this study. The point P2 was used to stimulate the bar in Fig. 5.15, and a directional coupler was positioned 30 cm away. Figure 5.22 illustrates how the produced numerical model is represented. Figure 5.23 compares numerical and experimental voltage waveforms and demonstrates the good agreement between the two types of voltage spectra . The staircase effect

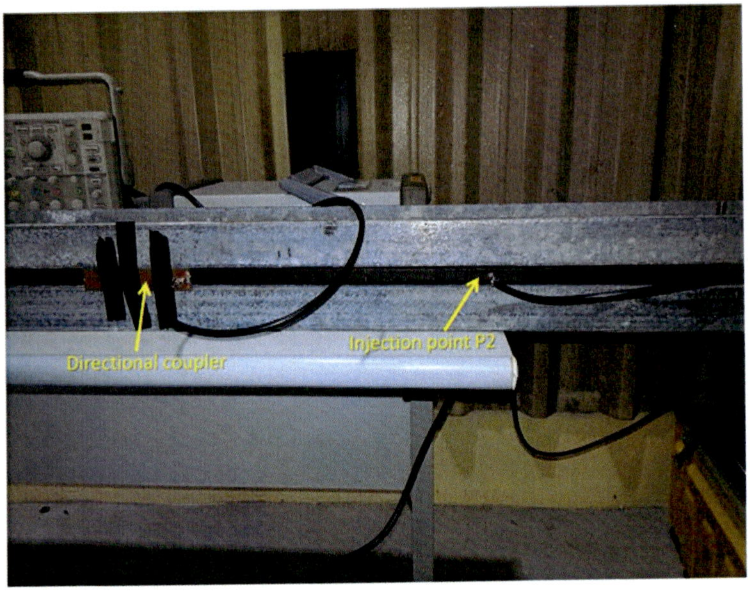

Fig. 5.21 Injection point P2 and directional coupler over the bar in experimental setup

Fig. 5.22 FDTD model of the hidrogenerator bar and directional coupler: numerical representation of experimental setup

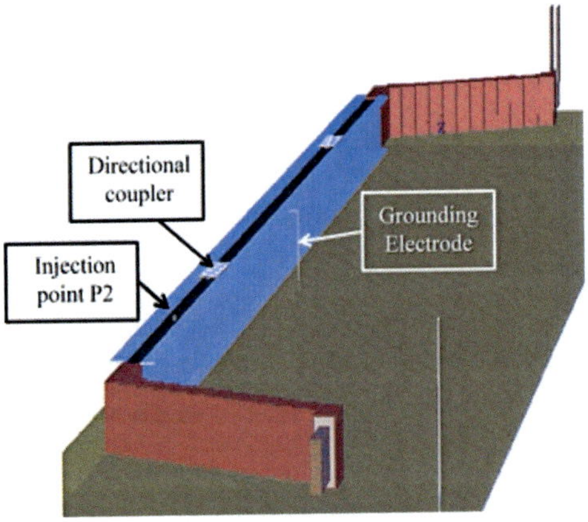

in the FDTD grid in the curved portions of the construction is responsible for the little discrepancy between the computational and experimental results.

Finally, every experimental finding is consistent with our theoretical and numerical analyses. Measurements reveal that resonances greatly depend on the geometrical characteristics of the bar, the placements of the discharge and sensor as well as on electromagnetic parameters, resulting in predictable spectrum patterns. This confirmation supports the methodological approach suggested in this work.

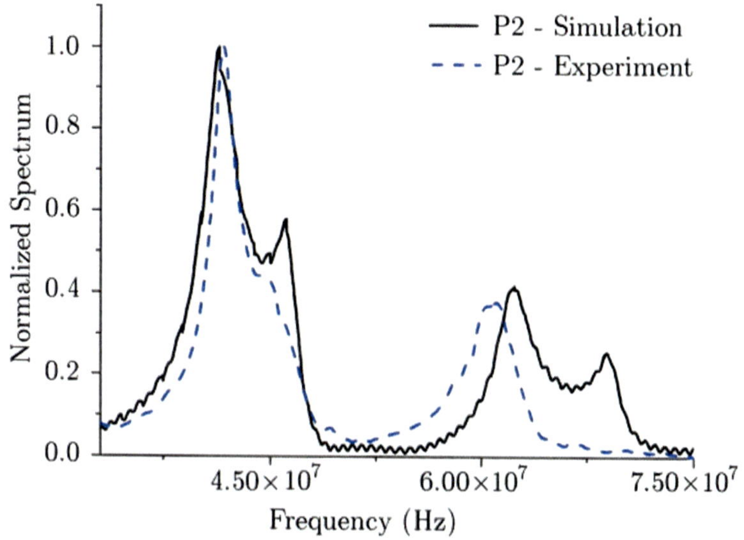

Fig. 5.23 Comparison of numerical and experimental voltage waveforms obtained with the directional coupler placed over the hidrogenerator bar

5.7 Final Remarks on PD Pinpointing

Based on spectral analysis, a new approach for localizing partial discharges was established. The electromagnetic process was modeled using the FDTD methodology. We were able to locate various partial discharges in one hydrogenerator coil as well as in two other coils that were linked to the first on each side. In conclusion, the method relies on identifying the resonance frequencies of the propagation channels between the positions of the transmitter (PD source) and the receiver (voltage sensor) inside the hydrogenerator coil. With this knowledge, the linearity of the Fourier transform and the Maxwell's equations, as well as the likelihood of discharges occurring as a function of coordinates, are taken into consideration to define a probability map of discharges occurrence points. This makes it possible to pinpoint areas where isolation issues occur.

In the scenario of a single coil, estimates of the location of one discharge with a maximum variance of 0.5 m were achieved in around 90% of simulations, while predictions of the positions of the second discharge with a maximum deviation of 0.5 m were made in 60% of simulations. The site of the first discharge was correctly calculated with a few centimeters of error from its precise position in the remaining 10% of simulations, while the location of the second discharge was determined with an uncertainty of 1.5 m.

The precise location of one of the partial discharges for two nearby coils could be estimated in roughly 81% of simulations, and in 47% of numerical trials, the second discharge position estimation had a variance of no more than 0.5 m.

The simulation results provided demonstrations of the success of the localization technique suggested in this chapter. Two important final comments on a real-world use of the approach are offered: (1) only the most hazardous PDs with high levels of amplitudes may be monitored and localized when there are many PDs in the system under investigation and (2) we made the assumption that the sensors in our study are perfect and do not cause any distortion to the PDs. The spectrum properties of the used sensors and electrical processing equipment should be taken into consideration while analyzing the real acquired signals.

References

1. Kim, H., Kong, T., Lee, S. B., et al. (2018). Experience with stator insulation testing and turn/phase insulation failures in the power generation industry. *IEEE Transactions on Industry Applications, 54*(3), 2225–2236.
2. Sadeghi, I., Ehya, H., Zarandi, R. N., Faiz, J., & Akmal, A. A. S. (2018). Condition monitoring of large electrical machine under partial discharge fault - a review. In *2018 international symposium on power electronics, electrical drives, automation and motion (SPEEDAM)* (pp. 216–223).
3. Mirzaei, H. R., Akbari, A., Gockenbach, E., Zanjani, M., & Miralikhani, K. (2013). A novel method for ultra- high-frequency partial discharge localization in power transformers using the particle swarm optimization algorithm. *IEEE Electrical Insulation Magazine, 29*, 26–39.
4. Biswas, S., Koley, C., Chatterjee, B., & Chakravorti, S. (2012). A methodology for identification and localization of partial discharge sources using optical sensors. *IEEE Transactions on Dielectrics and Electrical Insulation, 19*, 18–28.
5. Hoek, S. M., Kraetge, A., Kessler, O., & Broniecki, U. (2012). Time-based partial discharge localization in power transformers by combining acoustic and different electrical methods. In *International conference on condition monitoring and diagnosis* (pp. 289–292).
6. Kheirmand, A., Leijon, M., & Gubanski, S. M. (2004).Advances in online monitoring and localization of partial discharges in large rotating machines. *IEEE Transactions on Energy Conversion, 19*, 53–59.
7. Birlasekaran, S. (2003). Identification of the type of partial discharges in an operating 16kv/250 mva generator. In *Conference on electrical insulation and dielectric phenomena, annual report* (pp. 559–562).
8. Wang, W., Li, C. R., Li, W., Liu, L., Wang, Z., & Ding, L. (2001). Pattern recognition of single and composite partial discharge on generator stators. In *Annual report conference on electrical insulation and dielectric phenomena* (pp. 335–339).
9. IEEE Trial-Use Guide to the Measurement of Partial Discharges in Rotating Machinery. (2000). IEEE Std. 1434, Institute of Electrical and Electronics Engineers. IEEE.
10. Bozzo, R., Gemme, C., Guastavino, F., & Guerra, G. (1997). Localization of partial discharge sites on power generator bars by means of ultrasonic measurements. In *IEEE Instumentation and measurement technology conference* (pp. 658–663)
11. Wenzel, D., Borsi, H., & Gockenbach, E. (1994). Partial discharge recognition and localization on transformers via fuzzy logic. In *Conference record of the IEEE international symposium on electrical insulation* (pp. 233–236).
12. Luo, Y., Li, Z., & Wang, H. (2017). A review of online partial discharge measurement of generators of large generators. *Energies, 10*, 1–32.
13. Sadiku, M. (1995). *Elements of electromagnetics* (2nd ed.). Oxford University Press.
14. Ulaby, F. T. (2005). *Electromagnetics for engineers*. Prentice Hall.
15. Electrical Datasheet: U.H.E Coaracy Nunes. (1997). 1DH7139-3WF24-Z, Siemens.

16. Oliveira, R. M. S., & Sobrinho, C. (2009). Computational environment for simulating lightning strokes in a power substation by finite-difference time-domain method. *IEEE Transactions on Electromagnetic Compatibility, 51*(4), 995–1000.
17. Yee, K. (1966). Numerical solution of initial boundary value problems involving Maxwell's equations in isotropic media. *IEEE Transactions on Antennas and Propagation, 14*, 302–307.
18. Taflove, A., & Hagness, S. C. (2005). *Computational electrodynamics, the finite-difference time-domain method* (3rd ed.). Artech House.
19. Lesaint, O., Lebey, T., Dinculescu, S., Debruyne, H., & Petit, A. (2003). Propagation of fast pd signals within stator bars performance and limitations of a high frequency monitoring system. In *Proceedings of the 7th international conference on properties and applications of dielectric materials* (pp. 1112–1115).
20. Campbell, S. R., Stone, G. C., & Sedding, H. G. (1992). Application of pulse width analysis to partial discharge detection. In *IEEE international symposium on electrical insulation* (pp. 345–348).
21. Oliver, B. M. (1954). Directional electromagnetic couplers. *Proceedings of the IRE, 42*(11), 1686–1692.
22. Stone, G. C., Sedding, H. G., Fujimoto, N., & Braun, J. M. (1992). Practical implementation of ultrawideband partial discharge detectors. *IEEE Transactions on Electrical Insulation, 27*(1), 70–81.

Chapter 6
Digital Signal Processing Techniques Applied to Partial Discharge Monitoring and Classification

6.1 General Aspects of Partial Discharge Signals and Measurement

Partial discharges arrive at the recording system combined with the powerline signal, background noise, and other interferences. A simple model for such a composed signal is given by

$$v(t) = s(t) + d(t) + \eta(t), \tag{6.1}$$

where $v(t)$, $s(t)$, $d(t)$, and $\eta(t)$ denote the signal arriving at the sensor, the powerline signal, the PD, and a general term that represents all noise and disturbances, respectively; t indicates time throughout this chapter.

The powerline signal $s(t)$ is sinusoidal and can be expressed by

$$s(t) = A \sin(2\pi f_{\mathrm{N}} t + \phi), \tag{6.2}$$

where A represents the amplitude, f_{N} is the nominal frequency (either 50 Hz or 60 Hz), and ϕ denotes the phase of the powerline signal.

The term $\eta(t)$ is commonly considered as an association of different interference sources, like background noise and narrowband electromagnetic emissions. The background noise is the result of the thermal agitation of charged particles in the measuring system, being modeled as additive white noise with gaussian distribution (i. e., additive white gaussian noise or AWGN). The relative intensity of the background noise is characterized by the signal-to-noise ratio (SNR). In our context, the SNR, expressed in decibels, is defined as

$$SNR_{\mathrm{dB}} = 10 \log_{10} \frac{P_{\mathrm{PD}}}{P_{\mathrm{AWGN}}}, \tag{6.3}$$

V. Dmitriev et al., *Partial Discharges in Hydroelectric Generators*, Power Systems,
https://doi.org/10.1007/978-3-031-36604-8_6

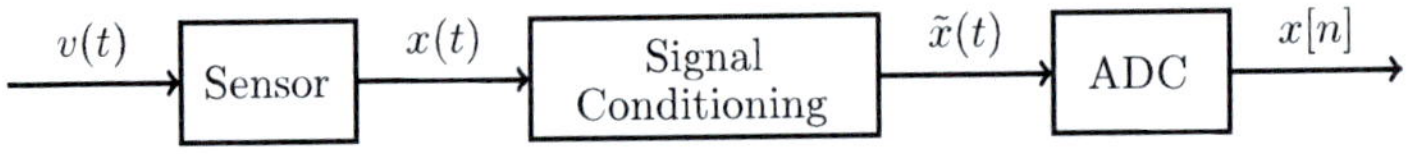

Fig. 6.1 Block diagram of an acquisition system for PD signal recording. The abbreviation ADC stands for *analog-to-digital converter*, while the other elements of the diagram are $v(t)$: input signal; $x(t)$: sensor output; $\tilde{x}(t)$ output of the conditioning circuit; and $x[n]$: digital signal, containing PD information for further analysis and classification

where P_{PD} and P_{AWGN} refer to the power of the PD signal and the power of the background noise, respectively.

In turn, the narrowband (a.k.a. harmonic) interference has two main sources: radio frequency emissions from wireless communication equipment and the operation of electronic switching devices. A valid model for this kind of interference is the generic amplitude modulated (AM) signal

$$\eta_{AM}(t) = \sum_{n=1}^{N} A_n \left(1 + \sum_{m=1}^{M} \mu_m \sin(2\pi f_m t)\right) \cos(2\pi f_n t), \tag{6.4}$$

where A_n and μ_m model the interference intensity; and the frequencies f_m and f_n determine the loci of the disturbance in the spectral domain.

A general acquisition apparatus to record PD signals comprises the following components (Fig. 6.1): sensor, signal conditioning circuit and analog-to-digital converter (ADC).

As we are interested in analyzing PDs, unwanted components in Eq. (6.1) must be opportunely removed or strongly attenuated. The fact that the powerline is a narrowband signal with fixed and known nominal frequency facilitates the task of filtering it out. For example, when using capacitive sensors, the powerline signal $s(t)$ can be eliminated from the measured signal $v(t)$ by the sensor itself, avoiding any additional processing step for this sake. The conditioning circuitry introduces other early modifications into the sensed signal, as a result of anti-aliasing procedures and operations for amplitude matching to ADC input specifications. That said, the signal at the ADC input $\tilde{x}(t)$ is mostly composed by the PD signal and some filtered noise components.

The output of the ADC module $x[n]$ is then a digital (i.e., sampled and quantized) signal that is stored for further processing and feature extraction. For simplicity, we consider a uniform sampling procedure (Eq. (6.5)) defined by

$$x[n] = \tilde{x}(nT), \tag{6.5}$$

with $T = 1/f_s$, where f_s is the sampling rate, and n is an integer that indicates the sample order.

Assuming the signal acquisition procedure is adequately performed (i.e., amplitude matching, anti-aliasing filtering, and sampling frequency are all correctly set), the digital signal $x[n]$ can be represented as

$$x[n] = d[n] + \eta[n], \tag{6.6}$$

where $d[n]$ and $\eta[n]$ are, respectively, digital versions of the PD signal and noise/interference term.

One should be aware that signal conditioning and analog-to-digital conversion are not perfect in practical implementations. Nevertheless, a rigorous treatment of the effects of non-ideal anti-aliasing filtering, and quantization noise due to finite-length binary representation is out of the scope of this text.

Precise definition of the sampling rate value depends mainly on a combination of two factors: the frequency range of the acquisition equipment and the spectral characteristics of PD signals. The first factor is addressed by the technical specifications IEC/TS 60034-27-2 [1], where four typical frequency ranges are identified:

1. Low frequency (LF) range:
    - Bandwidth: $\sim$ 1 MHz
    - Lower cut-off frequency: 100 s of kHz
    - Upper cut-off frequency: $<$ 3 MHz
2. High frequency (HF) range:
    - Bandwidth: from 3 MHz to 30 MHz
    - Lower cut-off frequency: $<$ 1 MHz
3. Very high frequency (VHF) range:
    - Bandwidth: few 100 s of MHz
    - Lower cut-off frequency: 30 MHz
    - Upper cut-off frequency: 300 MHz
4. Ultra high frequency (UHF) range:
    - Lower cut-off frequency: 300 MHz
    - Upper cut-off frequency: 3 GHz

Generally, the higher the working frequency, the better the SNR and the sensitivity for PD signals. However, with the increase of frequency, also the higher is the cost of ADC equipment and processors. Thus, there is a clear trade-off regarding frequency range, SNR levels and cost of the monitoring system [2]. A typical HF/VHF (high frequency/very high frequency) configuration for monitoring PD signals has an ADC sampling rate of few Giga-samples per second (e.g., from 2 to 5 GHz) with bandwidths of 100 s of MHz.

Figure 6.2 depicts the PD monitoring scenario in the frequency domain. Note that lower and upper cut-off frequencies of monitoring system (f_L and f_U, respectively) may vary according to the frequency range of the signal acquisition equipment. Also, note that the PD upper frequency of the PD signal arriving at the sensor (f_M) is lower than that of the PD signal at source location (f_0), because of the filtering effect along the propagation path between PD original site and sensor position. It is obvious that the PD monitoring system, due to its band-limited nature

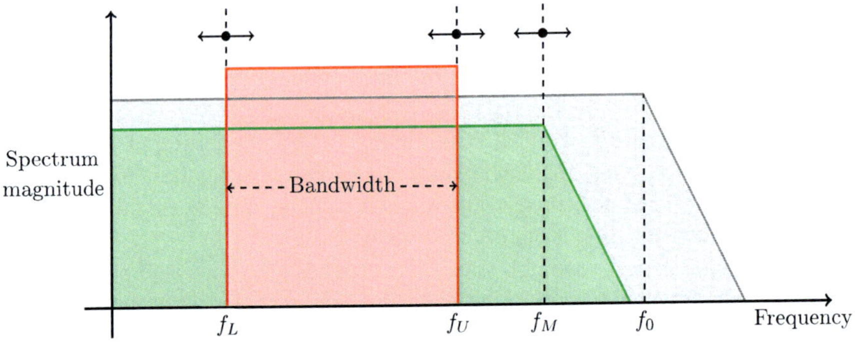

Fig. 6.2 Frequency bands in the context of PD monitoring. f_L : lower cut-off frequency of the monitoring system; f_U: upper cut-off frequency of the monitoring system; f_M: upper frequency of the PD arriving at the sensor; and f_0: upper frequency of the PD at source site. The arrows at the top of the figure indicate the key frequencies may vary depending on monitoring equipment (f_L and f_U) and on the propagation path from the PD site and sensor (f_M)

Table 6.1 Characteristics of different types of PD signals and noise/interference in the frequency domain [3]

Type	Fundamental spectral features
Internal PD	Relevant components around 50 MHz
Slot PD	Relevant components around 100 MHz
End-winding PD	250–500 MHz
Arcing PD	10–30 MHz
Vibration sparking PD	Relevant components around 6.7 MHz
AWGN	Wide band noise
RF	Narrowband interference with variant central frequency

cannot capture the whole PD phenomenon. The point here is to choose a working bandwidth that allows adequate PD characterization—good-to-high PD sensitivity and high SNR for those phenomena of interest—while keeping equipment cost within budget.

The second factor of interest—the spectral characteristics of the PD signals—must be taken into account in conjunction with available frequency information related to noise/interference components. This is necessary to implement effective separation of PD pulses from unwanted signals, a required condition for reliable extraction of PD features (e.g., pulse amplitude and duration, and powerline phase of PD occurrence). Partial discharge monitoring and classification are strongly dependent on the quality of such a feature extraction. Table 6.1 summarizes some standard information about general spectral attributes of different kinds of PD signals as well as of noise.

6.2 Partial Discharge Denoising

This section addresses techniques for denoising and source separation applied to the problem of PD monitoring. As discussed before (Sect. 6.1), PD pulses are recorded in conjunction with background noise and other interferences. The techniques discussed in this section comprise linear time-invariant filtering and thresholding of wavelet coefficients.The separation of PD pulses from noise is a crucial procedure to the quality of the PD monitoring system, due to its direct impact upon the PD feature extraction step.

6.2.1 Linear Time-Invariant Filtering

The very first procedure to analyze PDs consists in isolating the PD pulses from noise and interference. Even in the simplified model of Eq. (6.1), where the association of elements is additive, signal separation is not an easy task, due to spectral superposition. In practical situations, complete separation of $d(t)$ is not attainable and the best one can achieve is to decrease interference influence. Linear time-invariant (LTI) filtering is among the most straightforward strategies one can use to tackle this problem.

The LTI filter operation can be expressed in the frequency domain by the equation[1]

$$Y(\omega) = H(\omega)X(\omega), \tag{6.7}$$

where $Y(\omega)$ is the Fourier transform of the filter output, $H(\omega)$ represents the frequency response of the LTI filter, $X(\omega)$ is the Fourier transform of the input signal, and ω denotes frequency. Equation (6.7) emphasizes the interpretation of the LTI filter as a strategy to impose restrictions to the spectrum of the input signal.

The success of using LTI filtering for denoising is conditioned by the existence of a certain degree of spectral separation between signal and unwanted components. In this case, the general idea is designing a LTI filter of frequency response capable of removing or attenuating spectral regions associated with undesired signals, outputing PD pulses with shape and amplitude, among other features, closer to their actual values.

We assume, without loss of generality, that the powerline signal has been already filtered out either by the sensor or by the conditioning circuit (see Fig. 6.1). The denoising operation is performed in the digital domain, thus after the analog-to-digital conversion. Just to be clear, we reiterate we consider the signal $x[n]$ an additive association of $d[n]$ and $\eta[n]$—digitized versions of the PD signal and noise/disturbance—as stated by Eq. (6.6).

To catch a glimpse of what can be achieved by LTI filtering applied to PD denoising, we present some results by using simulated and experimental signals. To effectively test the performance of filtering strategies, we have implemented

[1] Additional discussion on linear time-invariant filtering is provided in Appendix D.

a simple signal generator in Python (available in book companion files), where PD pulses are simulated in conjunction with background noise and narrowband interference.

First, we consider a single PD pulse can generally be described by the *damped oscillatory pulse* (DOP) model [4, 5]:

$$p_k(t) = A_k \left[e^{-\alpha_k t} \cos(\omega_k t - \phi) - e^{-\beta_k t} \cos(\phi_k) \right] u(t), \tag{6.8}$$

where A_k, α_k, β_k, ω_k, and ϕ_k are the parameters of the model; t represents time; and $u(t)$ is the unitary step function.

In our simulator, the input signal $x(t)$ is a collection of random PD pulses, to which we add harmonic interference and background terms:

$$x(t) = \sum_{k=1}^{K} p_k(t - \tau_k) + \eta_{hi}(t) + \eta_{bg}(t), \tag{6.9}$$

where K denote the number of observed PD pulses; τ_k is the time shift of the k-th PD pulse; and $\eta_{hi}(t)$ and $\eta_{bg}(t)$ indicate the harmonic interference and background noise, respectively.

In this section, except for A_k, ω_k and τ_k that are randomly chosen for each $p_k(t)$ (Eq. (6.8)), model parameters are constant during simulation time. The parameters of next simulations (Table 6.2) are similar to those in other works on PD [4–6]. We invite the interested readers to access our simulator in the book companion files and try different setups to see how signal features are affected by their choices.

Table 6.2 Values of model parameters used in simulations: t_i and t_f correspond to the initial and final simulation time, respectively; f_s is the sampling frequency; $N(\mu, \sigma)$ and $U(a, b)$ are random values drawn from normal and uniform distributions, respectively. Other variable and symbol definitions are found in the text according to the given equation references

Parameter	Equation	Value
SNR	(6.3)	15 dB
N	(6.4)	1
A_1	(6.4)	1
ω_n	(6.4)	$2\pi\ 200 \times 10^6$ rad/s
ω_m	(6.4)	$2\pi\ 11 \times 10^3$ rad/s
μ_m	(6.4)	0.4
A_k	(6.8)	$N(1, 0.3)$
α_1	(6.8)	$20\,\text{Ms}^{-1}$
β_1	(6.8)	$250\,\text{Ms}^{-1}$
ϕ	(6.8)	0.12 rad
ω_k	(6.8)	$2\pi f_d \times 10^6$ rad/s
K	(6.9)	60
f_d	–	$U(1, 40)$ MHz
t_i	–	0 s
t_f	–	30 µs
τ_k	(6.9)	$U(t_i, t_f)$ µs
f_s	–	1 GHz

Figure 6.3a shows the whole ideal PD signal generated by the simulator in the time domain. We expanded the shaded area for a detailed view of a small group of PD pulses (Fig. 6.3b).

Next step, we added the background noise (AWGN) to the ideal PD signal (Fig. 6.4). The variance of the AWGN was chosen in a way to result a SNR of 15 dB. The same time slot as before is magnified, allowing a more detailed inspection (Fig. 6.4b).

To complete the synthesis of the observed signal, we include the harmonic interference term to the noisy PD signal (Fig. 6.5). Comparison between Figs. 6.3 and 6.5 gives us the sense of how difficult PD signal estimation can be if one looks exclusively at time domain.

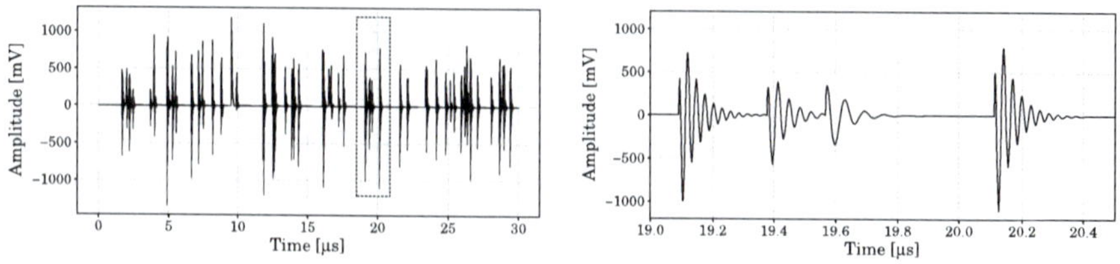

(a) (b)

Fig. 6.3 Simulated PD signal in the time domain: (**a**) complete signal; (**b**) magnification of the shaded area, showing a small group of PD pulses

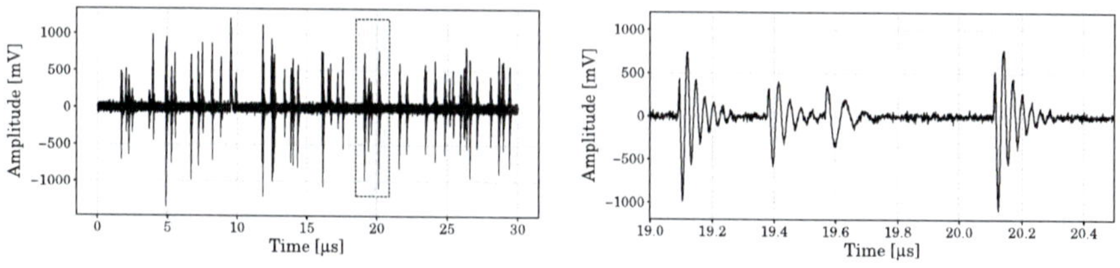

(a) (b)

Fig. 6.4 Simulated PD with background noise in the time domain ($SNR = 15\,\text{dB}$): (**a**) complete signal; (**b**) magnification of the shaded area

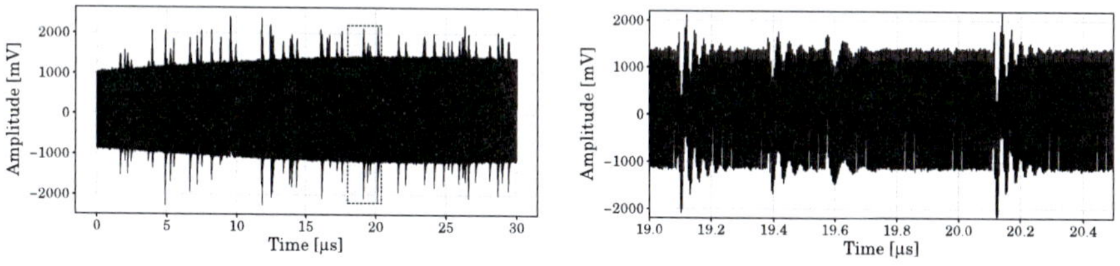

(a) (b)

Fig. 6.5 Simulated PD with background noise and harmonic interference in the time domain: (**a**) complete signal; (**b**) magnification of the shaded area

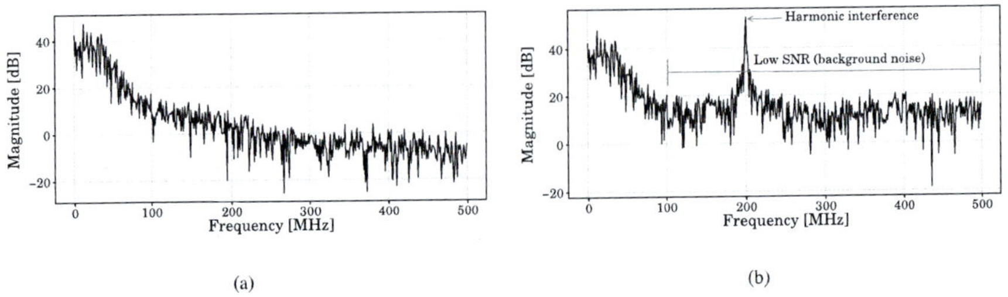

Fig. 6.6 Simulated PD in the frequency domain (magnitude): (**a**) ideal PD; (**b**) observed signal (PD + background noise + harmonic interference. Spectral Harmonic interference and background noise are both indicated

However, when we analyze the situation in the frequency domain, the task seems to be more tractable. Figure 6.6 shows the Fourier transform (magnitude) of both noiseless PD and observed signals. Comparison of magnitude levels between Fig. 6.6a and b shows it looks like relatively straightforward to figure out which frequency bands can be predominantly associated with PD pulses and noise/interference. Spikes in frequency domain indicate periodic signals. In our case, there is one very pronounced spike due to a harmonic interference with fundamental frequency of 200 MHz. The influence of the background noise can also be verified by the uplift in magnitude level for frequencies above 100 MHz.

When we recalculate the SNR considering the observed PD signal (the original PD with background and harmonic noises), we obtain the value of −15.87 dB, which indicates the energy of the interference largely exceeds of the PD signal.

Once we have identified which spectral region can be associated with PD signals—and thus targeted to be preserved after filtering—the next steps consist in the specification, designing and implementation of a LTI filter. For the synthetic signals at hand, we defined our filter as a being a bandpass with specifications given in Table 6.3. Essentially, our LTI filter aims to preserve frequency components between 5 MHz and 190 MHz. Keeping the same design specifications, we tried two different approaches: first, we design a finite impulse response (FIR) filter by using the windowing technique; and, in the second approach, an infinite impulse response (IIR) filter is implemented. The results for both designs are compared next.

Figure 6.7 shows the results for a FIR filter that uses a Hann window.[2] Figure 6.7a presents the Fourier transform magnitude of the input signal (observed signal), noise-free PD, output signal (filtered signal) as well as the frequency response of the FIR filter. In turn, Fig. 6.7b allow us a comparison between the observed (in gray), ideal PD (in black) e filter output (in red) signals for the same time slot defined previously. Note the delay between ideal PD and filtered pulses due to the high order of he FIR filter (the resulting order is 399).

[2]A discussion on the effect of different windows in given in Appendix D.

Table 6.3 Bandpass filter specifications for the synthetic signals. δ_p is the maximum variation of the frequency response magnitude of the filter in the pass-band; δ_s indicates the maximum gain of the frequency response magnitude of the filter in the stop-band; f_{pi} is the i-th pass-band frequency; and f_{si} denotes the i-th stop-band frequency

Design parameter	Value
δ_p	1 dB
δ_s	−30 dB
f_{s1}	0.001 MHz
f_{p1}	5.001 MHz
f_{p2}	190 MHz
f_{s2}	195 MHz

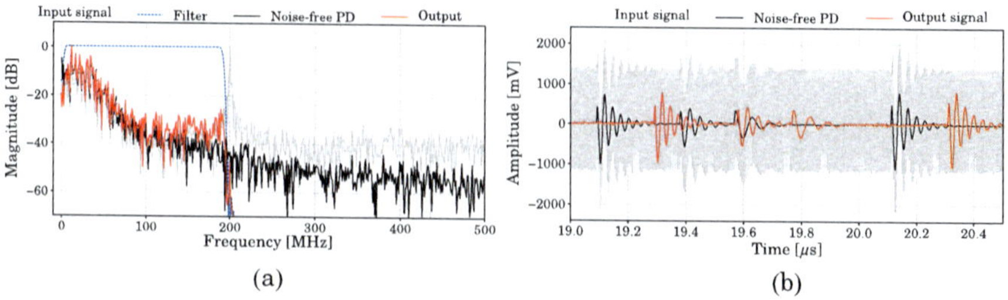

Fig. 6.7 Denoising of simulated PDs by using a FIR filter: (**a**) observed (input), ideal (noise-free) PD, filtered (output) signals and filter response in the frequency domain; (**b**) observed (input), ideal (noise-free) PD, filtered (output) in the time domain

To evaluate how effective is the filtering, we can calculate again the SNR, this time using the filtered and original PD signals. After compensating the delay of the filtered signal, we obtain for the SNR a value of 13.35 dB, an expressive improvement with regard to the observed signal.

As another example, we have designed an IIR filter that use Elliptic approximation (other approximations in Appendix D), whose results are shown in Fig. 6.8. Note the frequency response of the IIR filter differs significantly from the FIR, even though both were designed to meet the same specifications. As IIR filters allow feedback loops, their order tends to be much lower than FIR order (in this example, the resulting order is 15), meaning that IIR implementations require less computational resources (storage, arithmetic operations per sample, etc.) than FIR. However, with feedback loops, stability concerns are always present. Even if the delay of the filter output is not really an issue to be worried about—because it can be easily compensated a posteriori—compare how this characteristic is affected by filter order in both implementations. Independently of the approach, one can see the shape of the PD signal could fairly be recovered by both LTI filters for this synthetic case.

As done before, we calculate the SNR for the output of the IIR filter. The value obtained for this metric is 2.05 dB. Not so good as the SNR for the FIR filter, but a significant improvement anyway if ones considers the observed signal.

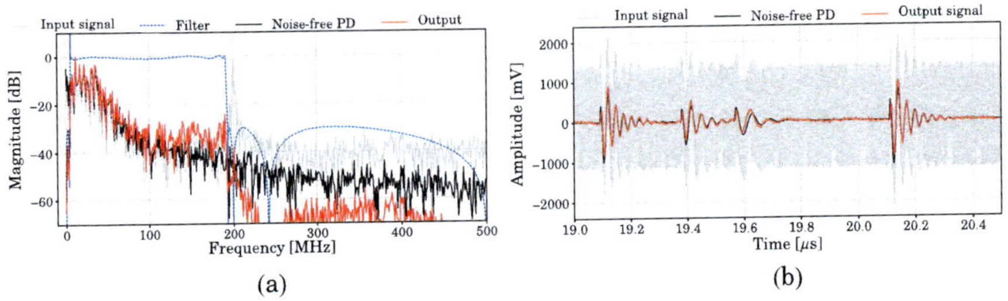

Fig. 6.8 Denoising of simulated PDs by using a IIR filter: (**a**) observed (input), ideal (noise-free) PD, filtered (output) signals and filter response in the frequency domain; (**b**) observed (input), ideal (noise-free) PD, filtered (output) in the time domain

Table 6.4 Bandpass filter specifications for the experimental signal

Design parameter	Value
δ_p	0.5 dB
δ_s	−30 dB
f_{s1}	1.0 MHz
f_{p1}	2.0 MHz
f_{p2}	7.0 MHz
f_{s2}	8.0 MHz

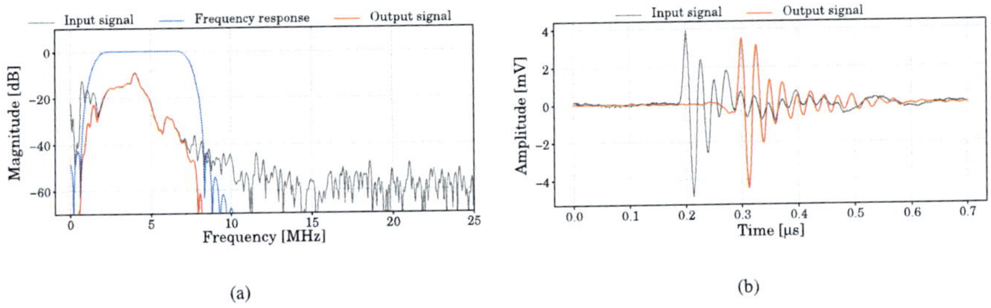

Fig. 6.9 Denoising of an experimental PD by using a FIR filter (experimental data): (**a**) observed (input), filtered (output) signal and filter response in the frequency domain; (**b**) observed (input) and filtered (output) signals in the time domain

Next, to give a sense of how these filters perform in practical situations, we present an example with an actual PD signal. This signal is just a single PD pulse, measured by a directional coupler from a non-impaired stator bar excited by a calibrator in controlled conditions (laboratory experiment). The sampling frequency of the data acquisition equipment is 250 MHz. As the characteristics of the actual PD signal differ from our previous simulated context, we have changed filter specifications in accordance with we can see in Table 6.4.

Like in the example with a synthetic PD signal, we design two LTI filters for comparison: a Hann FIR filter (order 500) and an Elliptic IIR filter (order 11). Results are given in Figs. 6.9 and 6.10 for FIR and IIR filters, respectively.

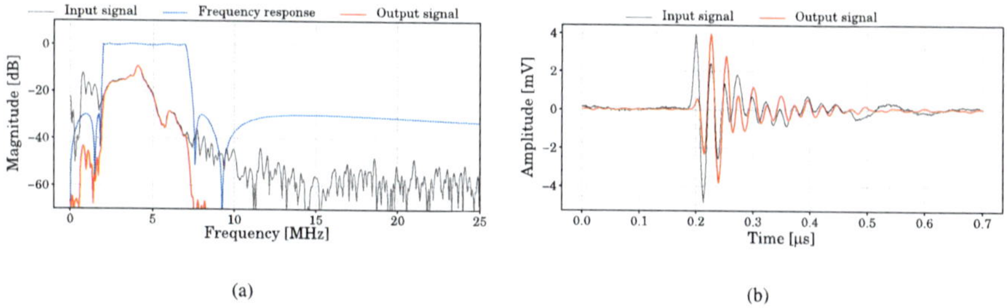

Fig. 6.10 Denoising of an experimental PD by using a IIR filter (experimental data): (**a**) observed (input), filtered (output) signals and filter response in the frequency domain; (**b**) observed (input) and filtered (output) signals in the time domain

The controlled conditions of the acquisition environment explains the low level of background noise in the input signal (Figs. 6.9b and 6.10b). In such a *quasi*-ideal situation, the filtering procedure surely does not exhibit impressive results, but such results can give us an idea of how real signals look like in time and frequency domains and how filtering operation affects them. Again, one should note the differences in the output delay due to the order of filters and the resulting pulse shape peculiarities in the time domain.

6.2.2 Denoising by Thresholding of Wavelet Coefficients

Wavelets have been successfully used for signal denoising and interference removal, specially when the signals of interest are non-stationary, or very localized either in time or frequency. Wavelets allow more flexibility than Fourier and windowed Fourier transforms regarding the definition of atoms used to analyze time-frequency information of signals [7]. To keep the focus on our main subject, we will just present a short theoretical introduction to wavelets in this section. Additional information and references on wavelets, however, are given in Appendix E.

That said, with wavelets, we can decompose functions in the same way we do with the Fourier transform or any other set of functions that play the role of basis for a Hermitian subspace. The concept is expressed mathematically in

$$f(t) = \sum_{j=-\infty}^{\infty} \sum_{k=-\infty}^{\infty} c_{j,k} \psi_{j,k}(t), \tag{6.10}$$

where $f(t)$ is the function we want to express by using wavelets; $\psi_{j,k}$ represents the scaled and translated version of the mother wavelet ψ; j and k denote the scaling and translating factors; and $c_{j,k}$ are the so-called wavelet coefficients.

Mother wavelets are designed in such a way that a set orthonormal functions can be obtained by translation and scaling as follows:

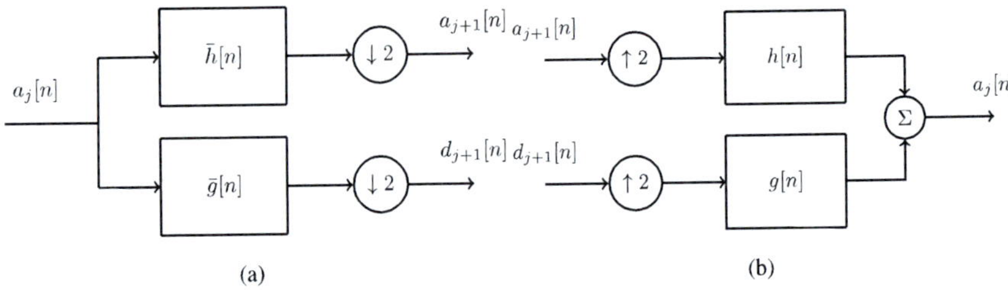

Fig. 6.11 Basic blocks to implement wavelet transformation by using filter banks: (**a**) one-level wavelet decomposition; (**b**) one-level wavelet reconstruction. The symbols ↓ 2 and ↑ 2 indicate the downsampling and upsampling operations by a factor of two, respectively

$$\psi_{j,k}(t) = 2^{j/2}\psi(2^j t - k) \quad j, k \in \mathbb{Z}. \tag{6.11}$$

In turn, the wavelet coefficients are determined by calculating the following inner-product:

$$c_{j,k} = \int_{-\infty}^{\infty} f(t)\psi_{j,k}(t)dt. \tag{6.12}$$

Practical implementations of the wavelet transform in the discrete domain uses filter banks in conjunction with resampling blocks [7, 8]. Figure 6.11 shows the basic block diagrams to perform one-level wavelet decomposition and reconstruction.

In Fig. 6.11a, $\bar{h}[n]$ is the impulse response of a low-pass filter and $\bar{g}[n]$ the impulse response of a high-pass filter. Those filters split the frequency information of a_j in two. The output of each filter is then downsampled by a factor of 2 (a_{j+1} and d_{j+1} have the half of the size of a_j); a_{j+1} and d_{j+1} are named *approximation* and *detail* components of a_j, respectively. The decomposition proceeds by repeating the processes with a_{j+1}, yielding a_{j+2} and d_{j+2}, and so on until the desired level of decomposition is attained. Wavelet reconstruction is then the inverse process: a_{j+1} and d_{j+1} are interpolated—upsampled by a factor of 2 and filtered—and added to recover a_j.

Al-geelani et al. [9] study the use of wavelets as a tool for denoising in the context of PD analysis. In particular, those authors discuss a technique that is the combination of a notch filter and Daubechies-2 wavelets, which has demonstrated good results to cope with noise caused by corona.

One of the most used approaches to reduce noise by using wavelets is the thresholding of wavelet coefficients. The idea behind is that the disturbing signal increases the original absolute values of the wavelet coefficients by some variable amount. So, the task is to identify which coefficients should be altered and to what extent. This way, the signal reconstructed by using such corrected wavelet coefficients will be closer to the noiseless signal. Thresholding techniques in general require the selection of the thresholding method and the definition of threshold values to be applied at each subband. Hard and soft thresholding are the predominant methods.

In *hard thresholding*, if the absolute value of a wavelet coefficient (d) is lower than the threshold λ, such a coefficient is zeroed:

$$\hat{d} = \begin{cases} d, & \text{for } |d| > \lambda, \\ 0, & \text{for } |d| \leq \lambda, \end{cases} \tag{6.13}$$

where λ is the threshold, d is the input wavelet coefficient, and $\hat{d}$ denotes the improved wavelet coefficient.

In *soft thresholding*, however, coefficients with absolute values higher than threshold are shrinkaged, i.e., re-scaled in a way to smooth the transition between coefficient values before and after thresholding [10]. Equation (6.14) describes the procedure.

$$\hat{d} = \begin{cases} d - \lambda, & \text{for } d > \lambda, \\ d + \lambda, & \text{for } d < -\lambda, \\ 0, & \text{for } |d| \leq \lambda. \end{cases} \tag{6.14}$$

Once the thresholding method is chosen—hard, soft or a variation of them—the determination of the threshold value is of central importance. It depends on the intended application and on the criterion used to assess the denoising effectiveness. Si et al. [11] address the problem of defining wavelet thresholds for PD denoising. Their mathematical framework is similar to Donoho's seminal paper [10] on denoising by shrinkage of wavelet coefficients. The difference relies in the use of a modified particle swarm algorithm to estimate optimized thresholds.

Next, we present a couple of examples of wavelet denoising applied to synthetic and actual PD signals. We have used the Daubechies-16 (mother wavelet) filter in conjunction with the soft thresholding approach. For the sake of simplicity, the threshold value is defined by the guidelines given in [10]:

$$\lambda_j = \frac{m_j}{0.6745}\sqrt{2\ln(n)}, \tag{6.15}$$

where λ_j is the optimum threshold value for decomposition level j; m_j denotes the median values of the wavelet coefficients at level j; and n is the length of the original signal.

Our first case is the synthetic signal in previous section, whose characteristics are defined in Table 6.2. Figure 6.12 demonstrates the wavelet filtering applied to improve a signal that is corrupted by background noise only. One can see from the spectral information (Fig. 6.12a) that the chosen strategy strongly attenuates frequency components of the input signal above 100 MHz. In time domain (Fig. 6.12b), the resulting effect is an output signal that is smoother when compared with the ideal PD signal. The SNR for the wavelet denoising procedure is of 6.86 dB. Nevertheless, the output signal is close enough to the original to allow a fair estimation of PD amplitude and phase of occurrence.

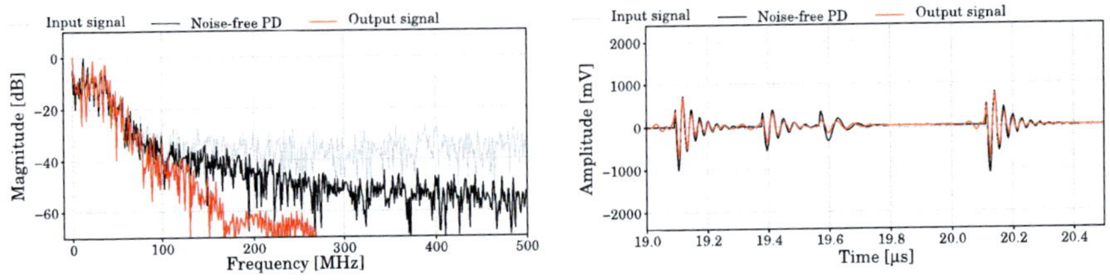

(a) (b)

Fig. 6.12 Denoising of simulated PDs by using wavelets: (**a**) PD with background noise (input), ideal (noise-free) PD, denoised signal by soft thresholding the wavelet coefficients in the frequency domain; (**b**) observed (input), ideal (noise-free) PD, and output in the time domain

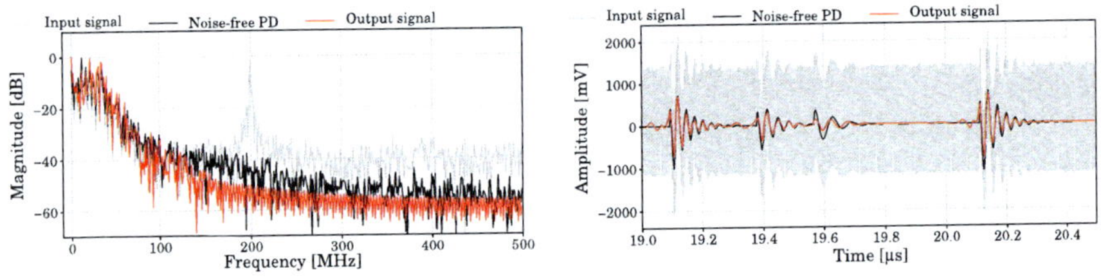

(a) (b)

Fig. 6.13 Denoising of simulated PDs by using wavelets: (**a**) PD with background noise and harmonic interference (input), ideal (noise-free) PD, denoised signal by soft thresholding the wavelet coefficients in the frequency domain; (**b**) observed (input), ideal (noise-free) PD, and output in the time domain

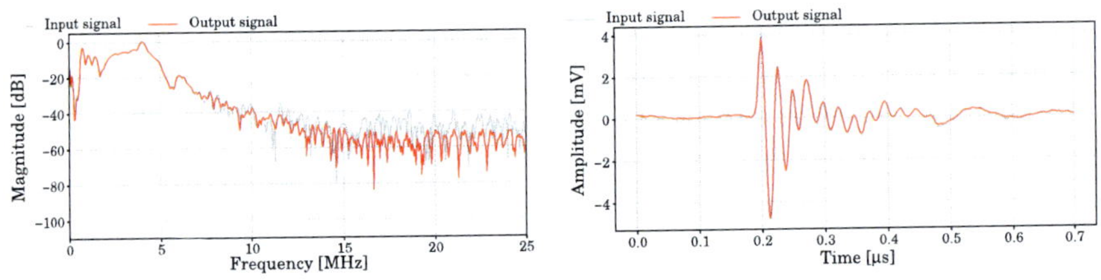

(a) (b)

Fig. 6.14 Denoising of an experimental PD by wavelet thresholding (experimental data): (**a**) observed (input), denoised (output) signal and filter response in the frequency domain; (**b**) observed (input) and denoised (output) signals in the time domain

The next example also includes the harmonic interference. Note the wavelet filter, in this case, has a softer impact in the high frequency components (Fig. 6.13a) than in the last example (Fig. 6.12a), because of the change in the wavelet thresholds due to the presence of the harmonic interference. Apparently, in time domain, however, such differences in high frequency components of the output do not influence significantly the final result (Figs. 6.12b and 6.13b).

We also applied the wavelet filter to the experimental PD signal presented in Sect. 6.2.1. Note the filtering has a slight effect in the high frequency region

Table 6.5 Values of SNR before and after denoising procedures for the synthetic example

Test signal	Comments	SNR
$x[n]$	Observed signal (without any processing)	−15.87 dB
$\hat{x}_{\text{fir}}[n]$	FIR filter output	13.35 dB
$\hat{x}_{\text{iir}}[n]$	IIR filter output	2.05 dB
$\hat{x}_{\text{wav}}[n]$	Wavelet filter output	6.86 dB

causing modest changes in the time domain (Fig. 6.14). Based on the SNR metric, a preliminary performance assessment of the denoising procedures presented so far is provided in Table 6.5. Note such results cannot be generalized as they were derived from one single simulated test signal.

6.3 Automatic Classification of Partial Discharges

There has been a growing interest in predictive maintenance (condition monitoring) within the electric power industry [12] for minimizing equipment downtime and associated costs. An important predictive maintenance technique for rotating machines is PD monitoring and analysis. This technique enables the evaluation of stator insulation condition. [13, 14]. Substantial PD levels are a relevant symptom of stator insulation aging in rotating machines, including hydrogenerators [15]. If not properly treated, PDs may progressively evolve to the point of disruptive failure in insulation [13, 16], causing great losses. PD monitoring is often performed online, mainly to avoid shutdown of the equipment [12]. Manned inspection of the large quantity of data generated by these measurements is unfeasible [13]. Thus, it is necessary to develop intelligent systems that automatically interpret these data. A key task for these systems is the identification of the kind of insulation defect causing the PDs. PD source diagnosis aids in planning appropriate maintenance measures for the examined apparatus, because each class of PD source has a particular evolution rate and imposes distinct risks to insulation [17, 18]. In literature, great effort has been made for PD recognition. In [19], four PD types are recognized in gas-insulated substations using decision trees, based on the relative concentrations of products of SF_6 gas decomposition due to PDs. Ma et al. [20] compare the recognition performances of different combinations of artificial intelligence algorithms and input features. They also propose a preliminary technique to recognize simultaneous PDs sources. Sinaga et al. [21] describe a PD classification methodology in power transformers, based on ultra-high frequency (UHF) signals. Statistical properties extracted from wavelet decompositions of UHF signals are used as input features for training an artificial neural network. In these references, as well as in most papers in literature, artificial PDs are generated in laboratory conditions to train and test proposed algorithms. The lower levels of noise and lesser ambiguity (simultaneous existing classes of PD sources) certainly result in superior recognition rates compared to a real-world application. Moreover, few works are dedicated to hydrogenerators, which are equipment of higher complexity.

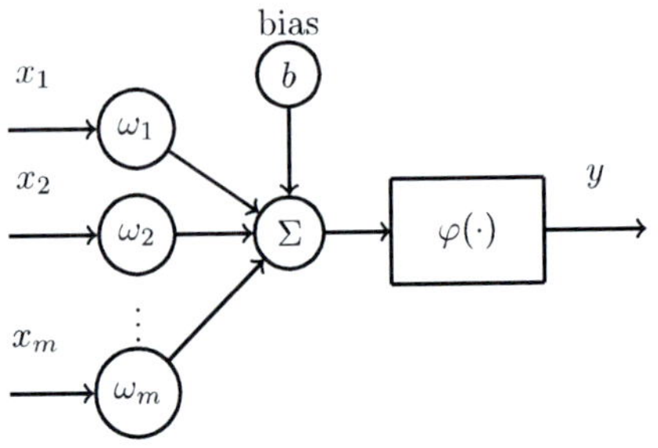

Fig. 6.15 Mathematical model of an artificial neuron

In this section, we present a methodology for automatic PD recognition in hydrogenerator stator bars. Actual PD samples were obtained by means of online measurements in hydrogenerators, operating in a real setting. Manual cleaning of the data is applied before feature extraction procedures. Such features serve as inputs for training and testing several classifiers based on artificial neural networks. Statistical tests are carried out to assess the performance and to select the best classifier.

6.3.1 Artificial Neural Networks

Artificial neural networks (ANNs) are a collection of interconnected processing units that is capable of *learning* relations between inputs and outputs. Their design mimics the human brain [22].

The basic processing unit of an ANN is the artificial neuron (Fig. 6.15). To calculate its output (y), the neuron applies the following mathematical operations to the inputs x_i:

$$y = \varphi\left(\sum_i \omega_i x_i + b\right), \tag{6.16}$$

where ω_i are the synaptic weights; b is the bias, which serves to increase or decrease the activation function input; and φ is the activation function, which limits the neuron output to a predetermined range [22].

In order to infer complex relations between inputs and outputs, multiple neurons are connected in several architectures. The architecture of interest in this work is the Multilayer Feedforward, shown in Fig. 6.16.

The neural network infers the relations between inputs and outputs by means of an iterative tuning of weights. In the supervised learning scheme, adopted in this work, the network is presented to input samples with known outputs during the training phase. Such a correspondence between inputs and outputs is generally obtained by hand labeling [22]. Once trained, a neural network can classify, with some level of error, any sample (set of inputs) consistent with any of the patterns to which it has been exposed during the training process. The goal is to automatically classify samples not present in the training dataset.

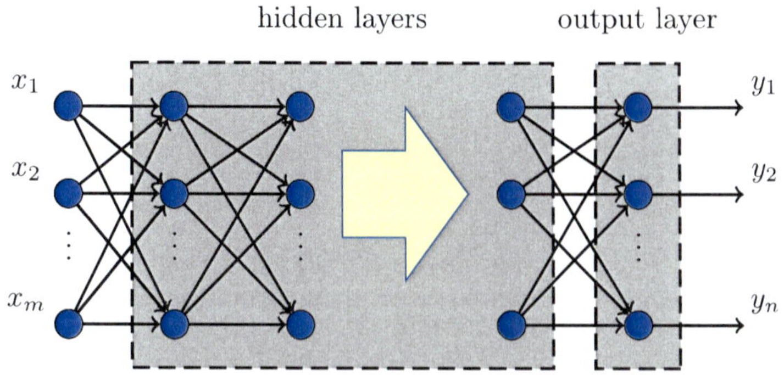

Fig. 6.16 Generic multilayer feedforward neural network

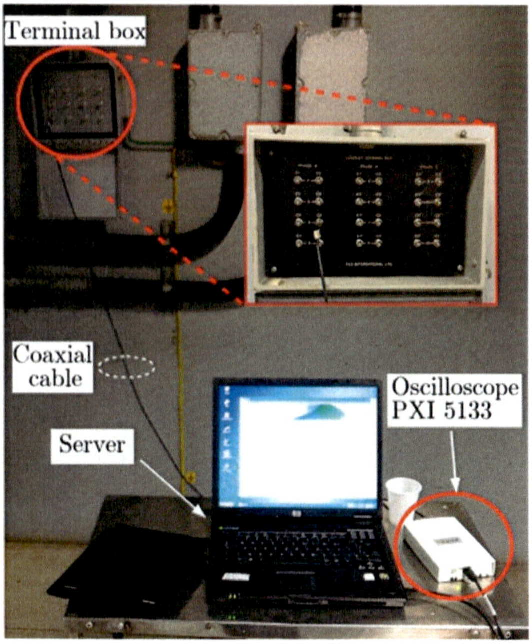

Fig. 6.17 Online PD measurement system in a hydrogenerator at Tucuruí power plant. Inset highlights the terminal box. Source: adapted from [23]

6.3.2 *Partial Discharge Dataset*

Our classifier was trained and tested with actual PRPDs obtained at the power plant of Tucuruí (northern Brazil). Figure 6.17 shows part of our experimental setup at Tucuruí.

Capacitive couplers are installed close to several stator windings of the monitored hydrogenerators. The output transient voltage signals were measured with the acquisition system called Instrumentation for Monitoring and Analysis of Partial Discharges (IMA-DP) [24]. This system applies several proprietary filters to the

Table 6.6 Distribution of PD sources in the dataset

Type	Number of examples
Gap discharge	331
Surface discharge	63
Slot discharge/Corona	113
Internal void	61

measured signal, and stores the information about the peaks of the PD signal detected during acquisition. From the stored peaks, IMA-DP builds PRPD diagrams with dimension 256 × 256.

The whole dataset contains 568 PRPDs measured in real hydrogenerators (Fig. 6.17). The collected samples were manually labeled by a human specialist among five of the most common PD sources described in IEC 60034-27-2 [1]: internal void, slot, corona, tracking and gap discharge (Table 6.6 and Fig. 6.18). Internal delamination was not considered because available samples yielded a certain degree of classification uncertainty: they can be easily misclassified as internal void, even by the human expert. The delamination type, which takes place between conductor and insulation, was also ignored due to the few samples in the dataset. Also, the samples of slot and corona were merged into a single class due to their similarities in the PRPD domain [18], reducing the problem complexity. Table 6.6 shows the distribution of PD sources in the dataset. It is clearly observed an imbalance among the classes as a result of the predominance of certain PD types (classes) in the investigated machines.

The PRPDs in the dataset present different degrees of difficulty for recognition. Some of the samples clearly belong to their respective classes, as shown in Fig. 6.18. Other samples are difficult to classify because they are ambiguous, exhibit high levels of noise, or result from multiple PD sources (Fig. 6.19). Automatic recognition of complex patterns is problematic [1], and their presence is expected due to the occurrence of multiple simultaneous PDs sources (of different classes) in rotating machines, and due to the intense noise and interference from power systems themselves [13].

6.3.3 The Proposed Classification Methodology

With the methodology presented here, one can perform automatic recognition of the primary (dominant) PD source using artificial neural networks. The complete flowchart of the methodology is shown by Fig. 6.20.

Three stages are considered: training, validation and testing. In the training and validation stages, the objective is to find the best neural network topology for the PD recognition task. Specifically, during training the optimal weights and biases of a neural network are calculated based on a subset of the dataset (the training set), whereas in validation the error over a second set acts as a stopping criterion

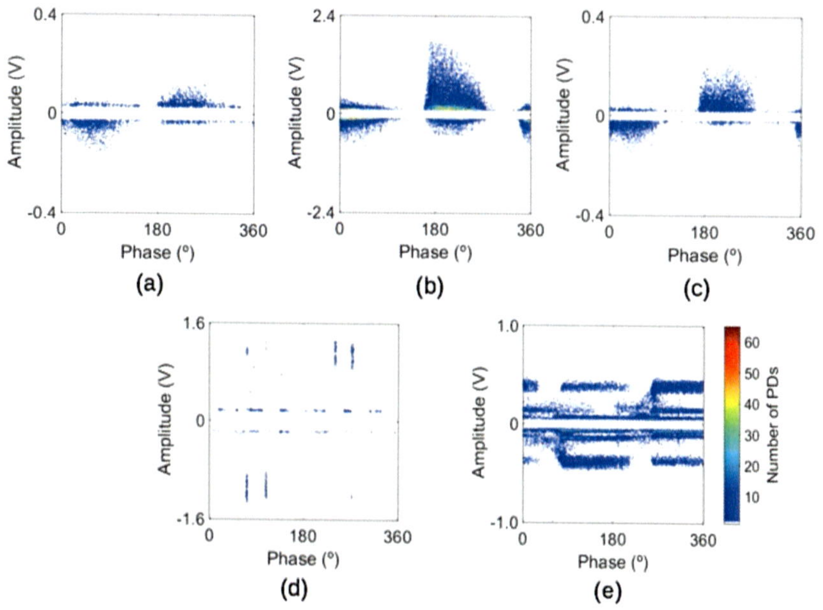

Fig. 6.18 Examples of PRPDs from the dataset. (**a**) Internal void. (**b**) Slot. (**c**) Corona. (**d**) Surface tracking. (**e**) Gap discharges. Source: [23]

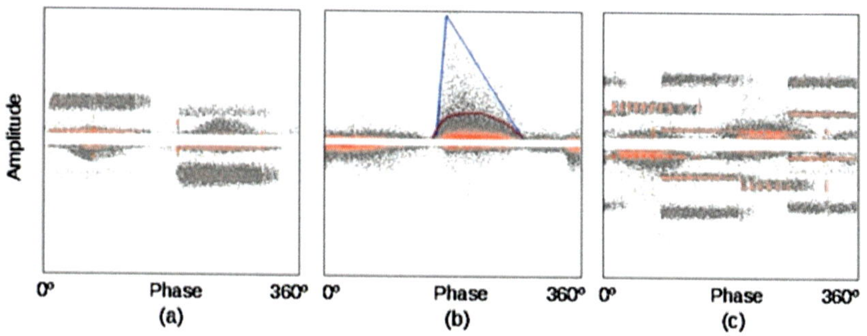

Fig. 6.19 Examples of complex patterns from the database: (**a**) gap discharges close to internal void clouds; (**b**) uncertainty between slot (triangular blue contour) and corona (rounded purple contour); (**c**) gap discharges superposed onto internal void, as well as strong noise. Source: [23]

of training to avoid overfitting [25]. In the test stage, samples used neither in training nor in validation steps are presented to the trained classifier to estimate its performance on unseen data.

The dataset is static: its samples have been previously labeled. In these conditions, it is plausible to treat the data with manual techniques in order to reduce the influence of features from non-dominant classes and/or excessively noise-contaminated examples that may degrade classifier learning. In all stages, the measured samples are subjected to the same preprocessing techniques for noise

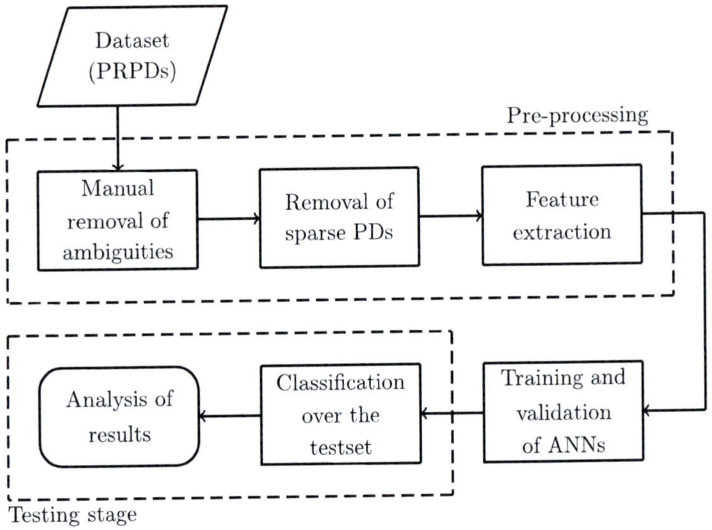

Fig. 6.20 Flowchart of the classification methodology. Source: adapted from [23]

treatment and extraction of input features, producing the data effectively presented to the ANNs.

Manual Removal of Class Ambiguities
This data cleaning technique consists of manually removing class ambiguities, i.e., PDs relative to sources apart from the PD source of interest. For example, in Fig. 6.19a, the source of interest can be considered to be internal void, while gap discharges can be judged to produce ambiguities. Removing this kind of ambiguity is necessary for the training process; otherwise the classifier would blend different classes, reducing the training effectiveness.

After carefully inspecting the database, all ambiguities clearly separable from the clouds of interest have been removed. However, it is not possible to eliminate ambiguities superposed onto the main clouds, as doing so would distort the PD pattern of interest. Clouds in PRPDs with superposed ambiguities were kept unchanged in the database as long as the ambiguities do not significantly change input features regarding the PD source of interest, such as shape and symmetry.

Each row of Fig. 6.21 contains the results of ambiguity removal applied to the patterns of Fig. 6.19a and c, depending on which PD source is considered to be of interest. In Fig. 6.21a, there is no overlapping between gap discharges and internal void clouds. Therefore, ambiguity removal of this kind of PRPD consists of completely removing the ambiguous PD source, as shown in Fig. 6.21b and c. In the PRPD of Fig. 6.21d, on the other hand, some gap discharges are superposed onto internal void. In this case, all ambiguities are removed except those superposed to the primary clouds, as shown by Fig. 6.21e and f. Notice that elimination of superposed PDs of different classes is not necessary for proper ANN training since

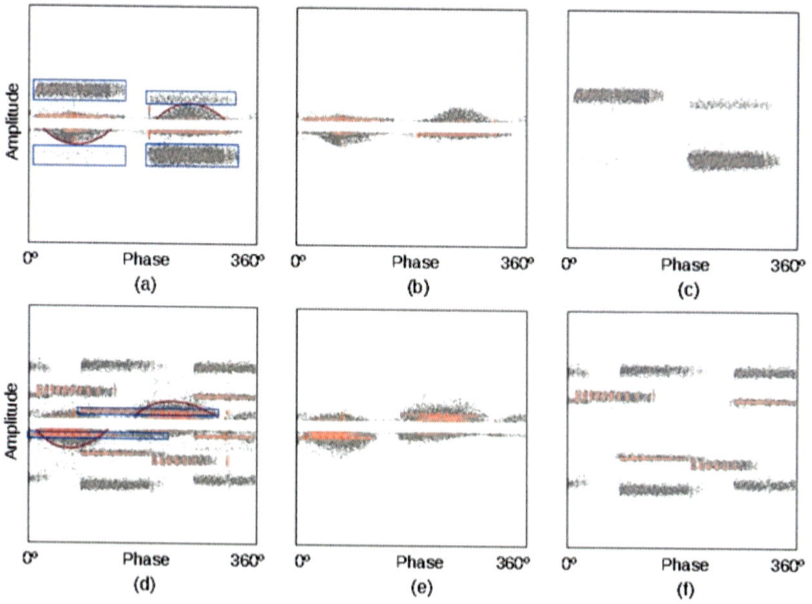

Fig. 6.21 Results of ambiguity removal applied to two measured PRPDs (**a**) and (**d**), considering that the PD source of interest is internal void (**b**) and (**e**) or gap discharges (**c**) and (**f**). Source: [23]

they do not significantly affect main features of PD clouds of interest, as depicted by the filtered PRPDs of Fig. 6.21e and f.

Removal of ambiguities is the only manual step in the methodology. It is automatically performed in other works using algorithms to separate multiple simultaneous PD sources, such as clustering in time-frequency maps [17]. However, this is not possible with PRPD maps because they are based solely on PD peaks.

Removal of Sparse Partial Discharges

Next, we use a morphological operator to eliminate isolated elements (PDs) in the PRPD patterns. Here, PRPDs are treated as binary images, whose pixels of value "1" correspond to non-zero PD counts (Fig. 6.22).

First, an observation region of size 5×5 is defined and centered on the pixel of interest. The filtering consists of zeroing center pixels that are the only nonzero element inside the observation region. Fig. 6.22 shows the results of such filtering applied to a PRPD of the dataset. The observation region slides across the PRPD until every pixel is evaluated.

The dimensions of the observation region were chosen empirically to keep a good balance between eliminating isolated PDs and preserving the main PD clouds (Fig. 6.22).

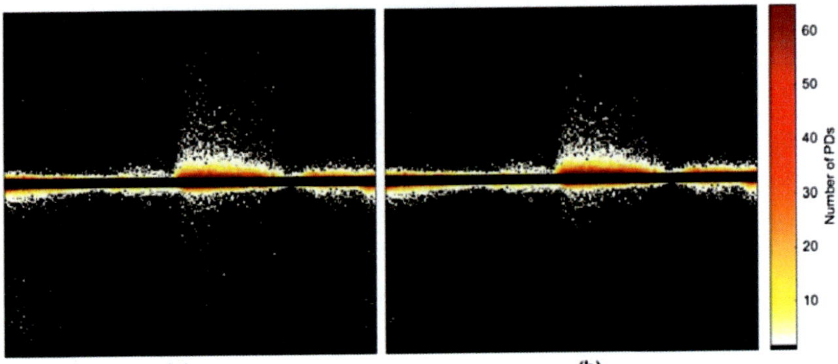

Fig. 6.22 PRPD (**a**) before and (**b**) after sparse PD removal by the pixel submatrix technique. Source: [23]

Feature Extraction
Each PRPD in our database is a 256 × 256 matrix, totaling 65,536 elements. Using all PRPD elements as input features would incur in a problem of very high dimensionality for a conventional ANN. In this situation, many features are of minor relevance, and a single point contributes very little to differentiate among classes [25]. Therefore, it is convenient to extract from the filtered samples a subset of features of more significance for performing the classification of the PD type.

The input features used in this work are novel. They are based on the projection of normalized PRPD counts onto the vertical (amplitude) and horizontal (phase) axes. The projections onto amplitude and phase were calculated for each polarity separately in order to properly characterize clouds of positive and negative PD amplitudes. This division is important to define, for example, the presence of symmetric or non-symmetric clouds or the shape of the clouds. These features can be determinant to properly perform PD classification [1, 18].

Given a PRPD map, the first step is to divide (normalize) all PD counts by the maximum counting found in the map, resulting in a $m \times n$ matrix M. Normalization eliminates the effect of random variables related to PD activity and measurement procedures, such as the severity of the insulation defect and its distance to the sensor, and time of acquisition. The projections onto the positive and negative amplitudes P_{a+} and P_{a-}, and onto phase in positive and negative polarities P_{f+} and P_{f-} are calculated using (6.17) and (6.18), respectively, which are given by

$$
\begin{aligned}
P_{a+}(i) &= \sum_{j=1}^{n} M(i, j), \\
P_{a-}(i) &= \sum_{j=1}^{n} M\left(i + \frac{m}{2}, j\right), \quad \text{with } i \in \left\{1, 2, \cdots, \frac{m}{2}\right\},
\end{aligned}
\tag{6.17}
$$

$$\begin{aligned} P_{f+}(j) &= \sum_{i=1}^{m/2} M(i, j), \\ P_{f-}(j) &= \sum_{i=1}^{m/2} M\left(i + \frac{m}{2}, j\right), \quad \text{with } j \in \{1, 2, \cdots, n\}, \end{aligned} \tag{6.18}$$

where $M(i, j)$ is the normalized quantity of PDs at magnitude i and phase j.

The amplitude range containing the relevant PRPD information is not fixed across samples. In order to obtain this range, let i_0 and i_1 be the highest absolute positive and negative amplitudes. The amplitude ranges are bounded by $\max(i_0, i_1)$ and $-\max(i_0, i_1)$, as illustrated by Fig. 6.23. Next, 64 points are interpolated from both amplitude projections within the amplitude range, and 64 points are interpolated from the phase projections in both polarities over the 60 Hz phase cycle. Figure 6.23 illustrates this feature extraction procedure.

The 64 interpolated points from each of the four projections are the data effectively presented to ANNs, totaling 256 input features. This represents a significant dimensionality reduction. Moreover, in Fig. 6.23 it can be observed that amplitude projections preserve the asymmetry between positive and negative PDs, while phase projections indicate the cloud shape and its positioning along the 60 Hz phase cycle. With the developed procedure, one can obtain features compatible with the PD characteristics mentioned in IEC 60034-27-2 [1] to describe each PD type in qualitative terms, indicating that the selected input features can, in fact, be used to classify hydrogenerator PDs.

Training and Validation of Neural Networks

Neural networks were trained in such a way to reduce the influence of ANN topology and of the following random variables: data sampling and initial weights. The topology (number of hidden layers and number of neurons per hidden layer) is a free parameter of neural networks, and its optimal configuration is not known beforehand. For this reason, ANNs of several topologies were trained. In order to mitigate potential performance inclinations due to the particular distribution of data that compose the training, validation and test sets, each topology was trained with several partitions of the data, using the strategy of stratified fourfold cross-validation (CV) [25], repeated 10 times. Each CV partition consists of selecting twofolds to form the training set, one for validation and the other for testing. The 12 possible partitions are considered for each repetition of the cross-validation procedure.

In a given partition, the training samples are presented to the ANN, and its weights and biases are updated iteratively in this work by means of the Scaled Conjugate Gradient Backpropagation algorithm (SCG) [26]. Then, the validation error is calculated. To avoid overfitting [25], this process is repeated until the validation error starts to increase. Once trained, the neural network classifies the test set, and the resulting error is used to estimate the error over unknown samples.

Initial weights have a major influence on training. The training of neural networks is a complex optimization problem. The goal is to find the combination

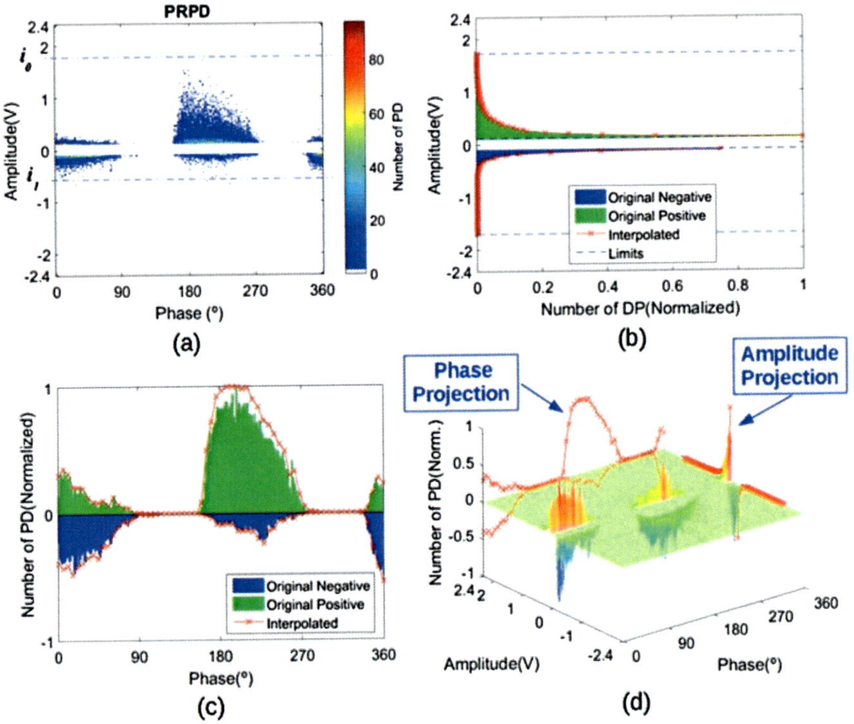

Fig. 6.23 Feature extraction. Input features are red points. (**a**) PRPD. (**b**) Projections onto amplitude. (**c**) Projections onto phase. (**d**) PRPD surface representation in 3-D space along with the amplitude and phase projections, which are perceived to be analogous to shadows of the PRPD surface. Source: [23]

of weights/biases (solution) that minimizes the difference (error function) between the ANN output and target classes. From a random guess, the SCG algorithm moves the solution iteratively in the direction of the negative of the error function gradient, converging to the closest point with zero gradient. Since the error function is generally non-convex and multimodal, gradient-based algorithms tend to get "stuck" at the closest local minimum. Initial solutions are totally random parameters, and the convergence to local optima is a characteristic of the training algorithm.

In order to mitigate the mentioned influence of the initial values of network weights on training, for each topology, 50 ANNs of different initial weights (defined randomly) were trained in each CV partition. Moreover, all ANNs have 256 input neurons, 4 output neurons, hyperbolic tangent activation function in hidden layers and softmax function in output layer.

6.3.4 Results and Discussion

All the trained ANNs were evaluated one by one in the following way: each ANN is executed over the test fold relative to the CV partition in which it was trained; for each of the C classes, it is calculated the true positive rate per class (T_S), which is equal to the rate of samples of a given class that were correctly classified. Mathematically, for the class i, we have

$$T_{Si} = \frac{n_{i,i}}{\sum_{j=1}^{C} n_{i,j}}, \tag{6.19}$$

where $n_{i,j}$ is the number of samples known to be of class i, but predicted to be of class j.

Then, we calculate the average μ_{Ts} and standard deviation σ_{Ts} of the C true positive rates per class as

$$\mu_{Ts} = \frac{1}{C} \sum_{i=1}^{C} T_{Si} \tag{6.20}$$

and

$$\sigma_{Ts} = \sqrt{\frac{1}{C} \sum_{i=1}^{C} (T_{Si} - \mu_{Ts})^2}. \tag{6.21}$$

The performance of each ANN is quantified by the novel metric δ, defined as

$$\delta = \mu_{Ts} + \frac{1}{1 + \sigma_{Ts}}. \tag{6.22}$$

As observed in Eq. (6.22), δ has two contributions: one increases with μ_{Ts} and the other decreases with σ_{Ts}. The first term is associated to the classifier's *average* performance across classes. The second term accounts for the variability of true positive rates per class. It distinguishes classifiers with similar μ_{Ts} by ranking higher those with more uniform success rates across classes (less biased). The idea of δ is to appropriately rank classifiers that perform similarly well at recognizing samples of different classes. Thus, a desirable classifier is that with high recognition rates for any class (high μ_{Ts}), with rates that do not differ much from one another (low σ_{Ts}).

Global success rate (rate of correctly classified samples relative to the size of the dataset), a typically used criterion for evaluating classifiers [25], is not suitable for problems with imbalanced datasets because it is biased towards the classes with most samples. Also, since δ is calculated over samples not used in

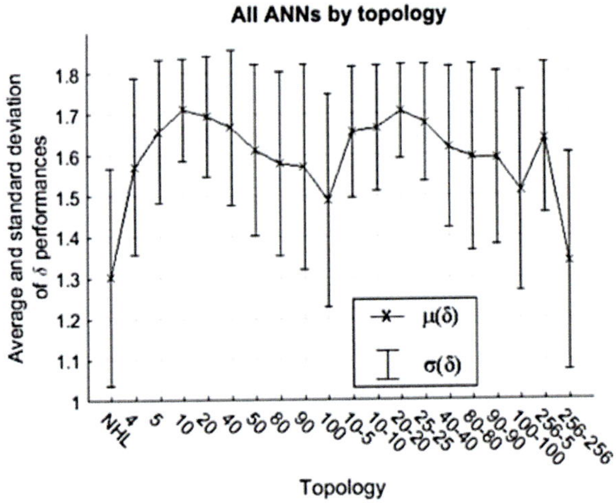

Fig. 6.24 Average (line) and standard deviation (vertical bars) of δ performances of all networks for each topology. Vertical bars are the size of two standard deviations. Source: [23]

training (i.e., we use for this calculation the test set), it estimates a classifier's generalization capabilities. It is worth mentioning that δ is a generic metric; it can be used to evaluate any classifier in any Machine Learning application involving the recognition of a single class per sample (multi-class, single-label problem).

In the following results, the neural network topology is expressed as the number of hidden neurons separated by dashes. NHL is the topology with *no hidden layers*: inputs are directly connected to output neurons. In addition, due to the fact that the initial weights are not related to the classification problem itself, their influences are minimized by showing results for the 25% best neural networks of each topology.

Figure 6.24 contains the average $\mu(\delta)$ and standard deviation $\sigma(\delta)$ of δ performances of all ANNs of each topology. Figure 6.25 shows these statistics for the 25% best ANNs of each topology. These data measure the influence of topology on performance, after varying initial weights and CV partitions.

Similar interpretation of δ is used to rank the different *topologies*: the best topologies are those whose networks present high average performances (high $\mu(\delta)$) and low standard deviation $\sigma(\delta)$ for the obtained values of δ. It is worth mentioning that $\mu(\delta)$ and $\sigma(\delta)$ are different than μ_{Ts} and σ_{Ts} from (6.20) and (6.21), respectively. The latter two indicate the variation of performance across classes, and are calculated for each network in order to calculate its performance δ. The former two, on the other hand, express the variation of δ performances of several networks for a given topology, and are used to select the most suitable topology (which is a free parameter of ANNs) for the problem at hand.

In Fig. 6.25, due to the reduced influence of random initialization of weights, one can observe that the average and standard deviation of δ are relatively constant over the range between topologies 50 and 256-5. Topologies 10, 20 and 40 presented

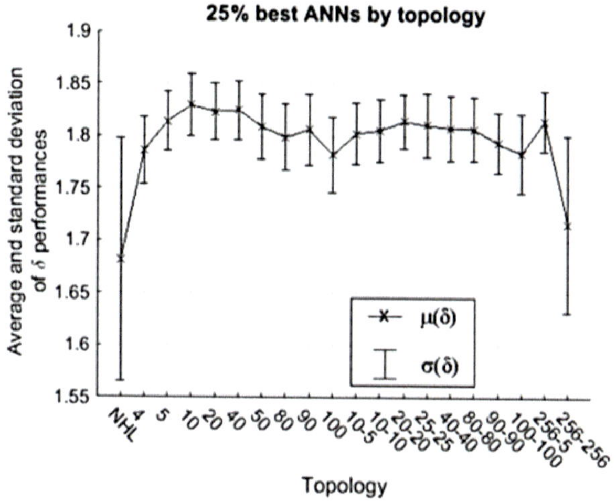

Fig. 6.25 Average (line) and standard deviation (vertical bars) of δ performances of the 25% best networks of each topology. Vertical bars are the size of two standard deviations. Source: [23]

larger $\mu(\delta)$ and similar $\sigma(\delta)$. The worst performance was observed for NHL (low average and large standard deviation of δs), indicating that the use of hidden layers is important to capture the non-linear relations between inputs and outputs. Similar observation was made in [20]. Bad results were also obtained for topology 256-256, probably because the number of samples was insufficient to train a network with so many weights and biases. Topology 10 is considered to be the most favorable on average, because it is the simplest configuration (low number of weights) of which most networks presented good generalization (high average and low dispersion of values of δ).

The curve in Fig. 6.24 presents higher variation in average δ across topologies, as well as much higher standard deviation due to random weight initialization influence. The poorer results are due to networks trained from inappropriate initial weights. Even with those differences, the best and worst topologies are the same as those mentioned above for Fig. 6.25, showing the representativeness of the 25% best networks in relation to all networks.

The next results are relative to the most favorable topology 10, and are illustrated by means of confusion matrices. A confusion matrix shows classification results intuitively [25]. Rows and columns are relative to true and predicted classes, respectively. Element (i, j) is equal to the number of samples known to be of class i classified as being of class j. Evidently, the correct classifications are evaluated by the elements in main diagonal.

To obtain a picture of average performance of all networks of topology 10, their confusion matrices over the testset were summed element-wise. Each element was calculated as the percent ratio relative to the sum in the respective row, resulting in the matrix shown in Table 6.7.

Table 6.7 Element-wise average confusion matrix of all ANNs of topology 10 over the testset (true positive rates are indicated in boldface). Source: adapted from [23]

	Predicted class			
True class	Internal void	Slot/corona	Tracking discharge	Gap discharge
Internal void	**78.6%**	7.53%	7.12%	6.71%
Slot/corona	5.22%	**92.8%**	0.88%	1.08%
Tracking discharge	4.38%	5.80%	**65.7%**	24.1%
Gap discharge	1.13%	0.14%	5.00%	**93.7%**

Table 6.8 Element-wise average confusion matrix of the 25% best ANNs of topology 10 over the testset (true positive rates are indicated in boldface). Source: adapted from [23]

	Predicted class			
True class	Internal void	Slot/corona	Tracking discharge	Gap discharge
Internal void	**85.8%**	3.63%	6.20%	4.37%
Slot/corona	4.83%	**93.9%**	0.54%	0.76%
Tracking discharge	2.24%	1.74%	**82.9%**	13.1%
Gap discharge	0.84%	0.10%	5.52%	**93.4%**

The same is performed in Table 6.8 for the 25% best networks of topology 10. In Table 6.8, the interpretation for element (1,3), for example, is that 6.20% of internal void samples were misclassified as tracking by the best networks, on average. The classes with the highest success rates were slot/corona and gap discharges, probably because these are the classes with the highest number of samples (Table 6.6). Smaller success rates were obtained for the other classes, indicating that the methodology can be worse at recognizing internal void and tracking patterns. Tracking samples are primarily misclassified as gap, while the inverse happens in a reduced degree. Confusion between tracking and gap discharges is reasonable, as these are the only classes formed by clouds of discharges far from the central region of PRPD. The same conclusions can be drawn from Table 6.7. The differences between these two figures—also caused by suboptimal initial weights—are lower recognition rates and consequently more frequent misclassifications.

The primary goal is to find an optimized network for trustworthy classifying PDs. Table 6.9 shows the confusion matrix of the best neural network (largest δ) of topology 10 over the test set. High recognition rates (of at least 94%) were obtained for all classes.

Graphical User Interface

A graphical user interface (GUI) was developed in order to display in an interactive way the results of PD classification and to automate the classification process. The classification is performed by a previously trained ANN by executing the test stage shown in Fig. 6.20.

The developed interface is shown in Fig. 6.26. It displays, from left to right, the PRPD of current sample, amplitude and phase projections, and pertinence probabilities of the current sample to the different classes. Pertinence probabilities

Table 6.9 Element-wise average confusion matrix of the 25% best ANNs of topology 10 over the testset (true positive rates are indicated in boldface). Source: adapted from [23]

	Predicted class				True positive
True class	Internal void	Slot/corona	Tracking discharge	Gap discharge	rate per class
Internal void	**15**	0	0	1	$\frac{15}{16} = 0.94$
Slot/corona	1	**28**	0	0	$\frac{28}{29} = 0.97$
Tracking discharge	0	0	**15**	1	$\frac{15}{16} = 0.94$
Gap discharge	1	0	4	**78**	$\frac{78}{83} = 0.94$

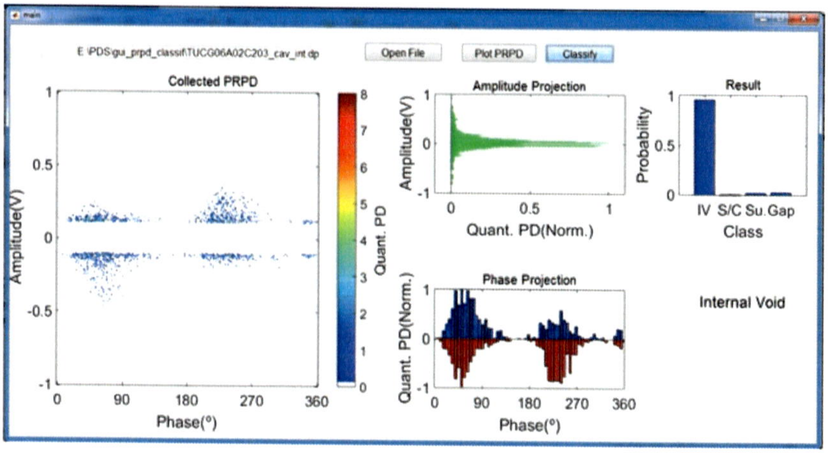

Fig. 6.26 The graphical user interface (GUI) developed to aid decision making. Pattern was correctly classified as internal void. Source: [23]

are proportional to the ANN outputs. The final PD classification—the class with the highest pertinence—is shown in the lower right corner.

Understanding the interface's intuitive indications does not require specialized operators, reducing training and operational costs and facilitating the application of the methodology in a real monitoring system by non-specialized human operators.

6.4 Supplementary Materials

We provide four Colab notebooks related to this chapter. The first notebook, *pd_simulator*, implements a partial discharge simulator based on the models described in Sects. 6.1 and 6.2. The user can set parameters for partial discharge pulses, background noise, and narrowband interference. The notebook provides basic graphical functionality for visualizing signals in both the time and frequency domains. Additionally, signals can be saved in a CSV file for further reference or processing.

The second and third notebooks (*pd_fir* and *pd_iir*) contain code for time-invariant filtering using Finite Impulse Response (FIR) and Infinite Impulse Response (IIR) filters. These notebooks are intended as a starting point for readers who wish to learn more about the subject. The notebooks provide visualization of impulse and frequency responses and allow users to store filter parameters.

The fourth notebook (*pd_wav*) provides examples of wavelet denoising using a thresholding approach. Users can select different wavelets, levels of resolution, and thresholding strategies.

To make the materials accessible to a general audience, we have hidden coding details in the *phd_book* module at the expense of flexibility and functionality. However, we provide the source code for all of our scripts to enable readers to explore the topics in more depth.

The Colab notebooks and some example data are available at

https://drive.google.com/drive/folders/1Y4PE50xru5qe0o9ZAh6HxFWFXDXk_Njg?usp=sharing

To use the content of *data* and *colab* folders, the readers must make a copy to their own Google Drive or download it to run locally with Jupyter.[3]

References

1. IEC. (2012). Rotating electrical machines-part 27-2: on-line partial discharge measurements on the stator winding insulation of rotating electrical machines. IEC/TS 60034-27-2. International Electrotechnical Commission, Geneva, Switzerland, Technical Report, IEC/TS 60034-27-2.
2. Stone, G. (2000). Importance of bandwidth in pd measurement in operating motors and generators. *IEEE Transactions on Dielectrics and Electrical Insulation*, *7*(1), 6–11.
3. Luo, Y., Li, Z., & Wang, H. (2017). A review of online partial discharge measurement of large generators. *Energies*, *10*(11).
4. Soltani, A. A., & El-Hag, A. (2021). A new radial basis function neural network-based method for denoising of partial discharge signals. *Measurement*, *172*.
5. Mota, H. O., Rocha, L. C. D., Salles, T. C. M., & Vasconcelos, F. H. (2011). Partial discharge signal denoising with spatially adaptive wavelet thresholding and support vector machines. *Electric Power Systems Research*, *81*(2), 644–659.
6. Chen, X., & Yang, Y. (2018). Analysis of the partial discharge of ultrasonic signals in large motor based on Hilbert-Huang transform. *Applied Acoustics*, *131*(2017), 165–173.
7. Mallat, S. (2009). *A wavelet tour of signal processing: The sparse way*. Academic Press.
8. Vetterli, M., & Kovačevic, J. (1995). *Wavelets and subband coding*. Prentice-Hall.
9. Al-geelani, N. A., Piah, M. A. M., & Bashir, N. (2015). A review on hybrid wavelet regrouping particle swarm optimization neural networks for characterization of partial discharge acoustic signals. *Renewable and Sustainable Energy Reviews*, *45*, 20–35.
10. Donoho, D. (1995). De-noising by soft-thresholding. *IEEE Transactions on Information Theory*, *41*(3), 613–627.

[3] https://jupyter.org/.

11. Si, W., Qin, B., Li, Q., & Liu, H. (2019). A novel adaptive wavelet threshold estimation based on hybrid particle swarm optimization for partial discharge signal denoising. *Optik, 181*, 175–184.
12. Stone, G. C. (2012). A perspective on online partial discharge monitoring for assessment of the condition of rotating machine stator winding insulation. *IEEE Electrical Insulation Magazine, 28*(5), 8–13.
13. Stone, G. C. (2005). Partial discharge diagnostics and electrical equipment insulation condition assessment. *IEEE Transactions on Dielectrics and Electrical Insulation, 12*(5), 891–904.
14. Stone, G. C., & Warren, V. (2006). Objective methods to interpret partial-discharge data on rotating-machine stator windings. *IEEE Transactions on Industry Applications, 42*(1), 195–200.
15. Malik, N., Al-Arainy, A., & Qureshi, M. (1998). *Electrical insulation in power systems*. Marcel Dekker.
16. Bartnikas, R. (2002). Partial discharges. their mechanism, detection and measurement. *IEEE Transactions on Dielectrics and Electrical Insulation, 9*(5), 763–808.
17. International Electrotechnical Commission. (2006). Rotating electrical machines - part 27: Off-line partial discharge measurements on the stator winding insulation of rotating electrical machines. IEC TS 60034-27. International Electrotechnical Commission.
18. Hudon, C., & Belec, M. (2005). Partial discharge signal interpretation for generator diagnostics. *IEEE Transactions on Dielectrics and Electrical Insulation, 12*(2), 297–319.
19. Tang, J., Liu, F., Meng, Q., Zhang, X., & Tao, J. (2012). Partial discharge recognition through an analysis of SF6 decomposition products part 2: Feature extraction and decision tree-based pattern recognition. *IEEE Transactions on Dielectrics and Electrical Insulation, 19*(1), 37–44.
20. Ma, H., Chan, J. C., Saha, T. K., & Ekanayake, C. (2013).Pattern recognition techniques and their applications for automatic classification of artificial partial discharge sources. *IEEE Transactions on Dielectrics and Electrical Insulation, 20*(2), 468–478.
21. Sinaga, H. H., Phung, B. T., & Blackburn, T. R. (2014). Recognition of single and multiple partial discharge sources in transformers based on ultra-high frequency signals. *IET Generation, Transmission & Distribution, 8*(1), 160–169.
22. Haykin, S. (2004). *Neural networks: A comprehensive foundation*. Prentice Hall.
23. Oliveira, R. M., Araújo, R. C., Barros, F. J., et al. (2017). A system based on artificial neural networks for automatic classification of hydro-generator stator windings partial discharges. *Journal of Microwaves, Optoelectronics and Electromagnetic Applications, 16*(3), 628–645.
24. Amorim, H. P., Jr., Carvalho, A. T., Oliveira Filho, O. B., Levy, A. S. F., & Sans, J. (2008). Instrumentation for monitoring and analysis of partial discharges based on modular architecture. In *2008 international conference on high voltage engineering and application* (pp. 596–599).
25. Witten, I., Frank, E., & Hall, M. (2011). *Data mining: Practical machine learning tools and techniques*. Morgan Kaufmann.
26. Møller, M. F. (1993). A scaled conjugate gradient algorithm for fast supervised learning. *Neural Network, 6*(4), 525–533.

Chapter 7
Partial Discharges and Ozone

The main method to monitor the condition of the stator insulation in hydrogenerators is the measurement of partial discharge pulses through capacitive couplers. However, partial discharges exposed to air also produce ozone, whose concentration can be measured through different types of sensors. This parameter can be valuable when assessing the condition of a machine, especially if the measurements are continuous. Since ozone concentration is affected by many variables, such as temperature, humidity, discharge voltage, airflow, etc, its value alone is not very meaningful, but it can be a good complement to typical partial discharge measurements, since it may reveal localized faults that would otherwise go unnoticed. It is also important to understand that very high ozone concentrations can chemically damage a generator, since it combines with atmospheric nitrogen and humidity creating nitric acid, which can corrode metals and degrade insulating materials.

7.1 Introduction

The name *ozone* derives from the Greek *ozein*, which means "to smell". Indeed, the ozone gas even in small concentrations has a distinct odour, which is part of what many people define as the "smell of rain". In 1785, Martin van Marum noticed that his static electricity apparatus produced a characteristic smell, which he attributed to electricity itself. Five decades later, Schoenbein [1] proposed that this smell was actually from a new substance and named it ozone. Only in 1865, Soret showed that this substance was in fact an allotrope of oxygen, whose molecule would have three atoms (O_3) [2].

Ozone is formed when an oxygen molecule (O_2) is broken into single oxygen atoms by the action of light or an electrical discharge, which then react with other oxygen molecules.

V. Dmitriev et al., *Partial Discharges in Hydroelectric Generators*, Power Systems,
https://doi.org/10.1007/978-3-031-36604-8_7

$$O_2 + electric\,discharge \rightarrow 2O, \tag{7.1}$$

or

$$O_2 + photon\,(\lambda < 240\,nm) \rightarrow 2O, \tag{7.2}$$

then:

$$2O + 2O_2 \rightarrow 2O_3. \tag{7.3}$$

Of course, the chemical reactions above are only an extreme oversimplification of what actually happens at atomic levels, because the electrical discharges excite O_2 molecules in various ways and the dissociation can be quickly followed by the recombination into O_2 instead of O_3. This is also not taking into account the vast amount of intermediate reactions that occur [3]. The actual rate of ozone formation is a parameter that is not easily calculated, especially outside of laboratory conditions.

Ozone is highly reactive and, for that reason, it is currently used in many kinds of applications, such as water treatment, sanitizing, food processing, etc. The main method to produce ozone commercially relies on corona discharges, which allow for concentrations of 3%–6% to be produced using nothing but atmospheric air. Other techniques exist, such as UV radiation (concentrations of up to 0.5%), electrolysis and even through radioactive rays, but none of them are nearly as efficient.

It is reasonable to expect, then, that such an efficient way of producing ozone would also be present unintentionally in most medium/high voltage equipment which are exposed to air and subject to electric discharges, including motors, generators, transmission lines, switchgear, etc. In fact, the production of ozone in electrical equipment as a "side effect" sometimes even becomes a health hazard for people working in the vicinity of such machines.

Inside air-cooled rotating machines, the electric discharges happen when the electric field is high enough to break down the gas molecules adjacent to the solid insulation. Another term used for this type of discharge is *corona*. The strength of the electric field (V_B) necessary to break down different types of gases is given by Paschen's law [4]:

$$V_B = \frac{Bpd}{ln[\frac{Apd}{ln(1/\gamma)}]}. \tag{7.4}$$

The parameters p and d are pressure and distance between the electrodes, respectively. Notice that what matters is their product, while A, B and γ are material properties of any given gas. In Fig. 7.1, from left to right, the breakdown voltage decreases as the pd product increases. This is because there are more molecules present to be ionised, freeing electrons to ionise other molecules, which will in turn ionise others, creating a self-sustained process which is the breakdown itself. However, after a certain minimum, the quantity of molecules starts to hinder this

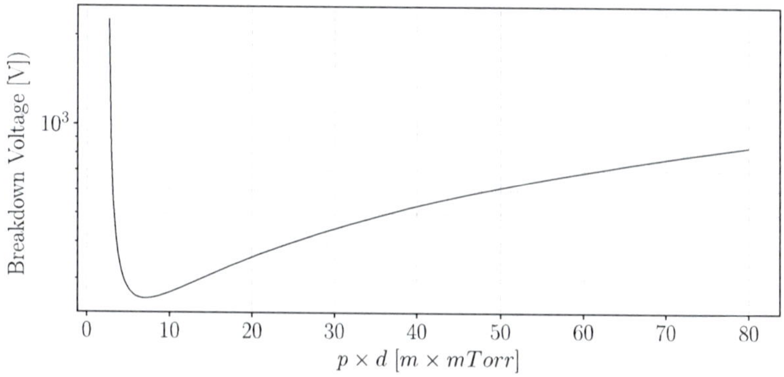

Fig. 7.1 Paschen's law plotted using parameters for air

process because electrons are colliding too frequently and therefore a larger voltage is necessary for the breakdown.

Assuming that a given machine exhibits this type of discharge and air is constantly being broken down, it is likely that ozone is also being produced at measurable quantities and its concentration will have a relationship with the intensity and number of discharges happening. This is why measuring ozone can help assess the general condition of the insulation, complementing the results provided by other electrical tests.

7.2 Formation of Ozone Due to Corona and PDs

In 1905, J. S. Highfield published a paper describing the observation of nitric acid inside a 3 MVA, 11 kV generator [5]. His hypothesis was that ozone created by electrical discharges was the cause for the subsequent formation of nitric acid, which was corroding the stator. This generator worked for a year before breaking down, and its insulation was of the highest quality (for the time, of course). Upon further inspection of the windings, it was observed that the turn insulation, made with cotton and resin, had been destroyed and replaced by a green substance. Also, the surface of the conductors had become rough. Some quantities of sulphuric acid, nitric acid and copper nitrate were found, giving more evidence that a chemical process had taken place. Highfield attributed this to the high humidity and performed laboratory experiments with the copper gauze wrapped in paper under a 10 kV potential difference. In 3 days, the paper was completely degraded and the copper was green, similar in appearance to the defective generator winding. He predicted that no generator in England could work effectively above 18 kV, due to the chemical degradation.

An independent work in 1907 [6] confirms that nitric acid was present in cotton interturn insulation, suggesting that for machines operating above 10 kV (phase voltage), precautions should be taken with respect to acid formation. Fleming [7] argues against this, saying that the insulation degradation occurred exclusively through the action of electrical discharges and not by chemical attack, adding that ozone could never be formed in measurable quantities at the working temperature of the generators.

Clearly, there was very little consensus about what exactly damaged the insulation of generators. It was only in 1911 that a detailed investigation on this subject was published. The paper *Chemical action in the windings of high-voltage machines* [8] should be read in full by all those who have an interest in rotating machines and partial discharges. It is the first work that clearly distinguishes between chemical degradation and the insulation wear caused by the constant action of electrical discharges (now called partial discharges). Giving a detailed explanation of the electric field gradients in the insulation, this work presents important conclusions:

- Chemical degradation is not perceived in machines operating below 6 kV (line voltage).
- Chemical degradation happens only in places exposed to air.
- Although ozone formation occurs in interfaces with air, chemical damage can happen in other regions of the machine, because the gas can move freely.
- The results of the chemical action in the insulation (whether by ozone, nitrous oxide or nitric acid) are oxidations.
- A failure by chemical degradation is almost always a short between turns.
- When the electric stress is too great, the degradation is not necessarily chemical. If there are two materials in the insulation with different values of permittivity, the material with smaller permittivity can be compromised by a large electrical stress without a full insulation breakdown.

In the following year, it was shown that electrical discharges also ionize the nitrogen gas present in air (N_2), which produces molecules NO_2 [9]. At that time, it was not known whether another product of this reaction is molecules NO.

Despite the strides towards a clearer comprehension of what happens inside a generator, until the 1940s there was still a lot of confusion around all these concepts. In 1943, it was shown by Von Cron that dielectric breakdown can happen after repeated discharges of small magnitude under constant electrical stress. This was an important result that helped understand how generators actually fail. After this, research on ozone inside electrical machines seemed to focus more on environmental and health implications.

A 1972 report [10] approaches the problem that transmission lines may pose to public health, considering the air quality standard of 0.08 ppm of ozone. The report determines that the ozone produced by corona is negligible considering the lowest detection limit of the equipment then used (2 ppm). A transmission line, however, is generally placed in open spaces, exposed to wind. For rotating machines, such as hydrogenerators, or even electrical cubicles, the situation is very different, as the gas can accumulate, leading to high concentrations.

The ozone formation through electrical discharges is not a simple process and its rate or efficiency depends on many variables. Two main parameters are mainly used: β which represent the number of molecules formed per unit charge transported and η which is the "energy efficiency" or the number of reactant molecules chemically changed per unit energy supplied to maintain the discharge [11]. The relationship between the two empirical rate coefficients for discharge reaction, β and η, can be traced in terms of the gas pressure, p, the electrode spacing, d, and the corresponding potential difference between the electrodes maintaining the discharge. Of course, the partial discharges happening in a machine stator during operation have a more complex behaviour, since the discharge path is not well defined and there are interfaces between different materials (mica, epoxy, steel, air, etc), not to mention the influence of the cooling air flow which has patterns that are far from trivial and will affect directly the rate coefficients.

Just as an example of this complexity, a simple experiment where an electrode discharged a DC current on two hollow needles separated by an equal distance shows that the current at constant voltage (or the total energy of the discharge) is not linearly proportional to the amount of ozone formed. This happens because if the energy is too high, thermal effects accelerate the destruction of ozone [12]. At the same time, it is possible to expect that greater electric fields tend to generate more ozone for a constant discharge current.

The decomposition of ozone is also another research subject [13], since it is a very reactive molecule. Some materials, such as copper, can act as catalysts in ozone decomposition. Humidity, temperature and presence of other gases further complicate the problem.

These factors are important to consider in ozone measurements. One implication is that the concentration of ozone alone can not reliably indicate the level of the surface PDs. For example, if two identical machines have their ozone concentrations measured at the same locations and found to have different values, the machine with higher concentrations will not always be in worse condition.

7.3 Problems Caused by Ozone Inside Electrical Machines

Occupational Safety and Health Administration (OSHA) regulations limit a worker's exposure to 0.1 ppm of ozone averaged on an 8-hour period. hydrogenerators can often generate as much as 1 ppm of ozone and upwards. There are examples of machines that need carbon filters in order to neutralize ozone and therefore lower the concentration that plant workers are exposed to. The accumulation of ozone is even more intense in the case of sealed machines (4 kV and above), such as motors rated for hazardous atmospheres, where the cooling air is not frequently renovated.

As discussed in the previous section, ozone can also become a hazard to the machine itself, damaging ferrous and rubber materials. Even though modern generators are made with materials much more reliable than the ones used in the early 1900s, some components are particularly susceptible to ozone damage: the iron stator core; brake ring; rotor shaft, hub, and rim laminations; air cooler fins and

Fig. 7.2 Example of chemical attack on the stator of a synchronous motor [15]. The white regions are not a contamination or powder, but a chemical change in the insulating material

gaskets; unpainted water piping and other exposed surfaces [14]. In some cases, the accumulated ozone reacts with moisture and atmospheric nitrogen gas producing nitric acid (H_2NO_3), which can quickly attack metallic and organic materials. The signs of chemical attack on mica-epoxy insulation are very easily spotted, because the surface becomes white, but not powdery.

The chemical attack on the insulation increases the amount of partial discharges in the surface, which in turn accelerate the generation of more ozone. On metallic parts, the chemical corrosion can lead to catastrophic failures. In the example on Fig. 7.2, the machine also had severe problems in the damper winding of the rotor, where some of the brazings were broken and corroded.

This failure mechanism can be easily spotted through visual inspections. Machines should be examined for red iron oxide deposits or loose powder on its surfaces, in the stator core air ducts, or in other areas in the stator housing. A borescope can be used to help with the inspection of areas otherwise inaccessible.

Additionally, stator end windings at the slot exits must be checked, because voltage stress grading paint or tape can become nonconductive as a result of electrical and thermal stresses. Sometimes, the interface between the grading coating and the semiconductive slot coating breaks down, creating a gap between the two coatings. This results in intermittent electrical discharges, increasing in frequency and intensity as the coating edges erode, leading to the formation of high quantities of ozone.

If a given machine has one or more of these symptoms, the severity of the degradation must be taken into consideration when deciding on corrective measures. In extreme cases, such as Fig. 7.2, the motor had to be fully rewound.

7.4 Ozone as a Diagnostic Method for PD

Ozone presence inside air-cooled generators and motors is always associated with the presence of partial discharges. Concentrations of just 0.1 ppm can already indicate severe PD activity. Stone et al. [16] also state that semiconductive coating deterioration can produce the most ozone.

As discussed previously in the chapter, the rate of ozone formation depends on a multitude of variables and not always provides a reliable assessment of the stator condition. It not always has a direct correlation with the total amount of partial discharges. When narrowing down to the partial discharges on the surfaces, then a better correlation might exist—namely slot PD, corona at the junction of the slot semiconductive coating with the voltage grading coating outside the slot, PD between adjacent coils in the end-winding, PD between adjacent connections and surface tracking. In any case, the trend of ozone concentrations inside a machine can be a valuable predictive maintenance tool when used in combination with other tests and parameters. It is possible to measure the ozone concentrations at periodic intervals with portable equipment or to monitor it continuously along other relevant variables.

There are studies where a possibility of even locating partial discharge spots by ozone measurements around the stator suggested. Figure 7.3 shows a situation where the general value all around the generator is comparable to ambient ozone, between 20 to 30 ppb, with the exception of two consecutive slots (15 and 16) where ozone concentration jump to above 50 ppb. This location contained the first bars of one parallel circuit of the three phases, which again are the most likely to give slot PD because of their higher potential. This suggests that the winding on this unit is in good general condition, with the exception of a very localized default. This may be a case of loose or deficient wedge in one or few slots [17].

A major challenge when using ozone concentrations to determine the approximate location of partial discharges is ventilation. Air flow severely affects concentrations. For instance, a lab experiment in a stator scale model showed that

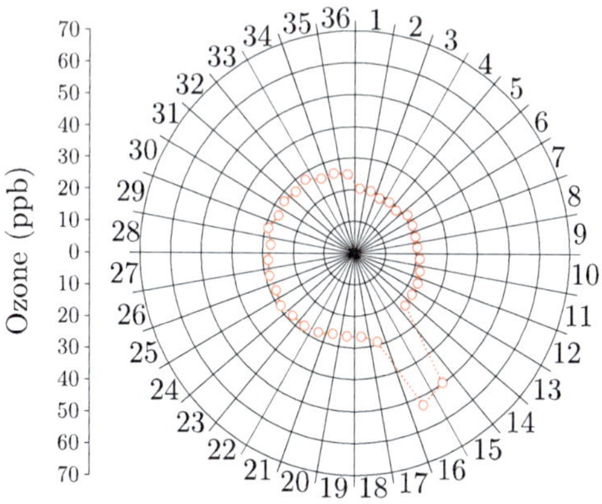

Fig. 7.3 Polar plot of ozone concentration around a hydrogenerator, adapted from [17]. In this case, a specific region of the stator is subject to higher partial discharge activity than the rest

concentrations of ozone of 30,000 ppb produced by a defect with static conditions and measured behind the stator core quickly dropped to 180 ppb at an air flow rate of 1.5 m/s. The higher the flow rate, the smaller the concentration. Moreover, single localized PD sites are unlikely to produce measurable quantities of ozone, especially if these sites are distant from the sensors or downwind with respect to cooling air flow.

Hence, the absolute levels of ozone inside a stator are probably not very meaningful (unless they are exceptionally high), but the comparison of measurements at different points as well as the determination of a baseline can be very useful to detect and confirm failure mechanisms that may be in action. Comparing the ozone concentrations with PD levels, machine load, voltage and temperature will certainly allow for a better diagnostic. If PD levels are increasing but ozone concentration remains stable, this could be an indication that PD is happening inside the insulation, for example.

There is another case where ozone measurements can help diagnosing surface PD. It is known that humidity severely affects the PD levels measured. In particular, high humidity levels tend to reduce the surface PD pulses measured through capacitive couplers [18]. This can be a bigger problem for off-line PD tests, when the machine is cool and exposed to higher humidity. In these cases, ozone measurements should be done as a complement to the traditional PD tests and will increase the reliability of the diagnostic. It is a good practice to record the temperature and humidity for every electrical test, since they usually influence heavily on the results.

In general, the ozone concentration can mostly be an indicative of surface discharges, such as:

1. Slot discharges.
2. Partial discharges due to vibration or displacement of bars.
3. Partial discharges due to contamination or tracking.
4. Partial discharges in the endwidings.
5. Partial discharges in the semiconductive/stress grading coatings.

For hydrogen-cooled machines, on-line ozone monitoring is not relevant since the quantities of oxygen are too small to form significant concentrations of ozone. If off-line tests are being done with the machine exposed to air, then it could be beneficial to record ozone concentrations with a portable equipment.

7.5 Measuring Ozone

There are many ways to measure concentration of ozone in air. Table 7.1 gives a general, non-exhaustive review of the main technologies in use. Each one has its own advantages and disadvantages, some of which will be discussed below.

For applications in electrical motors or generators, there are basically two main options:

Table 7.1 Comparison of different types of gas sensors [20]

Sensor type	Advantages	Disadvantages
Photo acoustic spectroscopy	High sensitivity; response time is fast; measurement is free from background noise; requires no reference as a result of noise	Selectivity is poor for photoacoustic system that utilises infrared light sources
Photo reductive gas sensor	Good sensitivity; short response time; inexpensive	Temperature requirement is high energy dissipation is high
Electro-chemical sensors	Portable; high sensitivity; inexpensive	Depletion of electrolyte when used for sensing high ozone concentrations requires frequent maintenance
Metal-oxide ozone sensors	Broad range of applications	High temperature requirements of detectors, implying in high energy consumption, high cost and fabrication/size limitations
Solid state	Consumes less energy; good sensitivity; fast response time; inexpensive; lightweight	Characteristic activity is high film sensor thickness requirement is large when applied for ozone sensing
Chemiluminescence	Fast response time	Requires to be calibrated within every hour and is not absolute
Optical spectroscopy	Accurate; lightweight; fast response time; rapid and direct means of sensing gases with good cross-sensitivity; requires no consumables either for calibration or operation; anti-EMI interference; excellent electrical insulativity; suitability of long-distance online measurements	Gas sample must be able to absorb, emit or scatter transmitted light rays in specific regions of the spectrum expensive large in size

1. Electro-chemical sensors, which are cheap but typically have smaller detection ranges and a life of 2–3 years (although there are manufacturers claiming up to 10 years [19]).
2. Optical spectroscopy, which have large detection ranges, good precision and long life span but are very expensive. This technology is also more sensitive to dust or other contaminants.

7.5.1 Electrochemical Sensors

Electrochemical sensors are a class of sensors where an electrode acts as transducer element, as shown in Fig. 7.4. This type of sensor first appeared in the 1950s, applied to industrial oxygen monitoring. Health and safety regulations started to establish limits for exposure to hazardous substances and Leland C. Clark proposed a concept for oxygen sensing consisting of two electrodes inside a cell with a membrane permeable to oxygen [21]. Upon chemical reduction in one electrode, an electrical current appears, which is proportional to the oxygen concentration in the sample. The same principle was applied to other types of gases. Since they rely on a chemical reaction, the sensor has a limited lifetime. Nowadays, there is usually a third electrode for reference purposes, improving signal-to-noise ratio, response time and stability.

Electro-chemical sensors are very cheap and versatile, since their electrical output can immediately be sent into signal conditioners, PLCs and similar equipment. The main limitation of this type of sensor is their dynamic range. Common values are 0–10 ppm or 0–20 ppm. For comparison purposes, optical sensors can easily measure 0–200 ppm with accuracy of 2%.

For the purpose of monitoring the condition of a hydrogenerator, however, the range 0–20 ppm is acceptable, since typical values recorded in peer-reviewed literature are below 1 ppm. Note that the sensor, in particular of the electrochemical type, will not be installed on the stator itself, but rather somewhere near it, for example in cooling air ducts, near the radiators around the machine, etc. This means that the concentration at the measuring point is expected to be much lower than in the immediate vicinity of the surface discharges. A big advantage of this type of sensor is that the signal generated is intrinsically electrical, which facilitates its conditioning and amplification. Other highlights are the possibility of miniaturization, relative low cost and fairly short response time. Electrochemical sensors can be classified by the electrical parameter that is changed in presence of the analyte.

The main types of sensors are presented in Table 7.1.

1. Potentiometric (electric potential V).
2. Conductometric or impedimetric (resistance or impedance Ω).
3. Amperometric (current I).

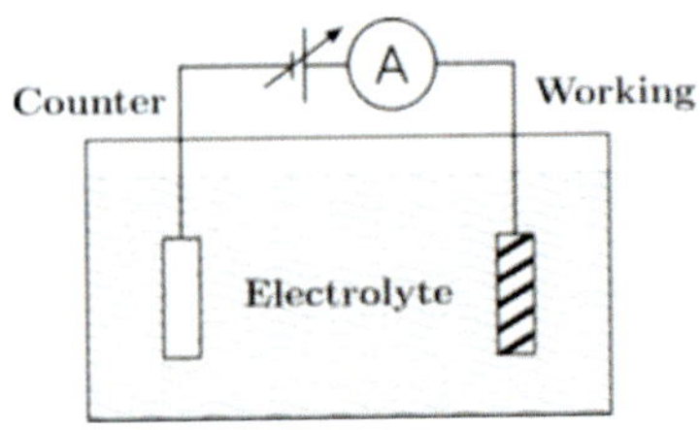

Fig. 7.4 General structure of an electrochemical gas sensor. The analyte diffuses through the electrolyte and reaches the electrodes, where a chemical reaction occurs

4. Coulometric (electric charge Q).
5. Capacitance (C).

With regards to ozone electrochemical sensors commercially available, the large majority are amperometric or impedimetric.

7.5.1.1 Amperometric Sensors

In its most basic configuration, amperometric sensors have 2 electrodes: one functional electrode and a counter-electrode. A constant reference voltage is applied between them and the presence of O_3 causes a reduction reaction in the surface of the functional electrode, at the same time that an oxidation occurs in the counter-electrode. This generates an electrical current that is proportional to the number of molecules of analyte that are present.

Modern amperometric sensors also have a third reference electrode, whose function is to keep the functional electrode at an electrochemical potential as constant as possible, even with changes in temperature, pressure, humidity or presence of contaminants.

Considering an ideal linear amperometric sensor, the gas concentration can be modeled by:

$$Gas\ concentration\ (ppb) = \frac{(WE_i - WE_0) - (RE_i - RE_0)}{Sensitivity}, \tag{7.5}$$

where WE and RE are, respectively, the changes in voltage at the **W**orking and **R**eference electrodes. Sensitivity, in this case, would have as unit V/ppb, even though the original signal is an electrical current. Common materials for the electrodes are silver, gold and platinum, which can be manufactured in various substrates, such as silicon [22] or even polymers [23].

One of the major advantages of amperometric sensors is the ability to operate at room temperature, with low energy consumption (typically in the order of tens of mW) and reduced sizes, allowing for battery operation, for example. Common measuring ranges for ozone are between 50 ppb to a few hundred ppm.

This type of sensor, however, is subject to quick degradation, usually lasting 1–2 years. This is due to the measuring process itself. For this reason, amperometric sensors are usually not very suitable for measuring high concentrations of ozone ($>$50 ppm). Also, amperometric sensors can have high cross-sensitivity with NO_2 [24]. In some applications, this characteristic can invalidate O_3 measurements altogether.

7.5.1.2 Impedimetric Sensors

Impedimetric or resistive sensors have a sensitive area whose resistance (or impedance) changes in contact with the analyte. In the same way as amperometric sensors, this change can be caused by redox reactions. In the case of ozone, these sensors are mostly made of metal oxides. For the reaction with ozone to occur, the material must be activated, which means reaching specific thresholds of conductivity, electrical mobility, charge carriers and chemical kinetics. This can be achieved with heat or light.

In the case of heated sensors, there is a resistive element that keeps the sensitive electrode at high temperatures (100–500°C). There is some confusion regarding how these sensors are named, because some of them are labeled as HMOS (heated metal oxide sensors), as if they were not electrochemical sensors. This is likely a marketing strategy, because the broader definition still applies. Even in peer-reviewed works, some authors place HMOS sensors in a separate category (see Table 7.1 for example). Their main disadvantage is the high power consumption. Due to the high working temperatures, theses sensors may be subject to instabilities and they can't be used in some places, such as areas with explosion risks. Measurements may also be affected by humidity present in the sample.

When light is used to activate the sensitive material (either visible or UV light), the sensor is often called photo reductive or photostimulated. It is also possible to find them labeled as "room temperature metal oxide sensors". Even with light excitation, some of these sensors may also require a minimal operation temperature.

A study with 143 amperometric and metal oxide sensors continuously collecting data for 5 months mentions that HMOS sensors are not as effective for measuring concentration peaks [25]. Although HMOS sensors can be more selective and have lower detection limits, they are also slower [26].

7.5.2 *Optical Sensors*

There are two main types of optical sensors: optical absorption (or photometric) and chemiluminescent (or photoluminescent). The photometric type is more widely used to measure ozone concentrations, especially when precision is a priority.

7.5.2.1 Optical Absorption Sensors

The O_3 molecule has a peak of absorption in the UV region, called Hartley band. This peak occurs at $\lambda = 253.7$ nm, as shown in Fig. 7.5

This peak can be used to determine the ozone concentration in a very precise way, using the Beer-Lambert law[28]:

$$A = \log_{10} \frac{I_0}{I} = L\epsilon c. \tag{7.6}$$

The absorbance (A) is given by the logarithmic ratio between the intensity of incident light (I_0) and the light intensity after crossing a medium (I). This is

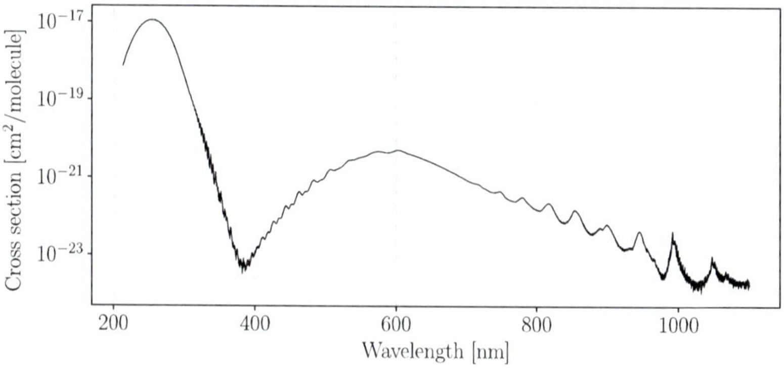

Fig. 7.5 Ozone absorption cross-section for 200 nm < λ < 1100 nm [27]. The peak at 253.7 nm is the Hartley band

Fig. 7.6 Diagram of a photometric ozone sensor. Source: adapted from [29]

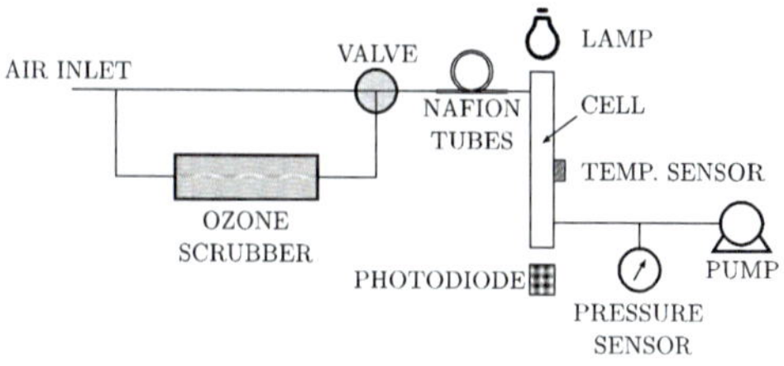

equivalent to the length of the medium multiplied by the molar absorbance (ϵ) and concentration (c).

Typically, photometric sensors have a scrubber system that allows all ozone to be eliminated from a sample. The sample first goes through the scrubber and has its absorbance measured. Then, a new sample is routed to the measuring cell without going through the scrubber. The measurements are compared and the real O_3 concentration can be determined. Figure 7.6 shows this process. In this device, there is also a Nafion tube for moisture removal, as well as pressure and temperature sensors that allow for a more precise volumetric concentration measurement.

This technology has a series of advantages, such as precision and stability. With an adequate design, ozone sensors of this type can detect levels below 10 ppb with small margins of error. The most noticeable downside of UV absorption technology is price, since it is more expensive than heated metal oxide, solid-state or electrochemical technologies. For some applications, the reliability of UV absorption can be worth it, especially when using a portable unit to take regular measurements across many machines.

Specifically, for hydrogenerators, there are systems provided by manufacturers that can measure ozone concentrations around the stator in real time continuously. This is done by placing tubes at each measurement point which suck in air to a manifold. The manifold sequentially directs the sampled gas of each tube to a measurement chamber where the concentration of ozone is evaluated and logged into the monitoring device [30].

The setup shown on Fig. 7.7 has a few disadvantages, for instance the air has to travel through a fairly long path to reach the actual measurement chamber, which means that changes in the O3 concentrations can happen. Also, the manifold has valves that open and close to switch between the different air inlets. These mechanical moving parts are more subject to maintenance issues and replacements.

Another type of optical sensor uses the same technology (UV absorption), but instead of bringing the air to a measurement chamber, it sends light through fiber optics, which then bounces off of a mirror inside an enclosure where the sampled air is. The light then comes back through another fiber into the spectrometer, where the UV absorption is measured and further correlated with the ozone concentration.

In the example from Fig. 7.8, the sensor described had a measurement range of 0.1–10 ppm and was intended for MV cubicles, but the same principle would apply for rotating machines. The advantages are that the air being sampled finds itself closer to the source of ozone (discharge). Also, the design of this sensor can be completely dielectric (fiber optics, lenses and mirrors), which means it could theoretically be placed as close to the stator as desired, or even glued to the endwindings, for example, similar to optical endwinding vibration sensors commercially available. The fiber optics would then be connected to an optical multiplexer or multi-channel spectrometer that would give the ozone concentration, eliminating the need for moving parts.

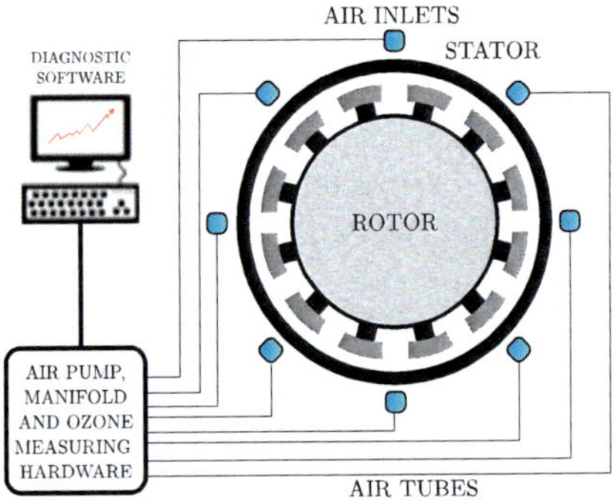

Fig. 7.7 Example of commercial system with air inlets and central measuring hardware

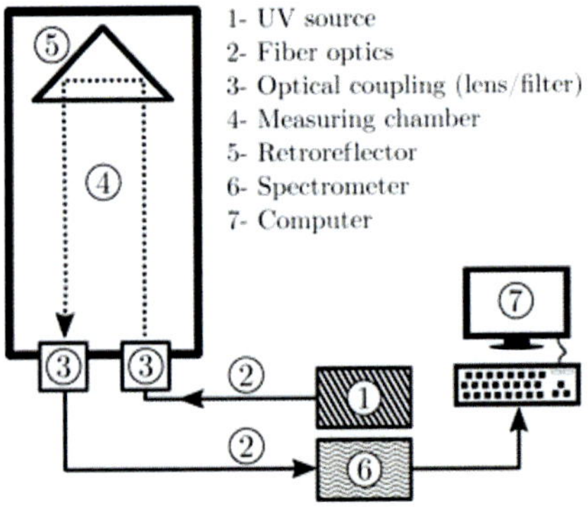

Fig. 7.8 UV absorption ozone sensor with fiber optics. Source: adapted from [31]

7.5.2.2 Chemiluminescent Sensors

Chemiluminescence is light emission caused by a chemical reaction. This technique uses a known reagent which is placed in contact with the analyte. The light emitted goes through a photomultiplier, and the signal amplitude is proportional to the analyte concentration. This method is used in laboratories, because it has extraordinary sensitivity and dynamics range. At the same time, it demands stable conditions, has very high costs and slow response times (20–30 min). Common uses are automobile emission certifications, combustion efficiency evaluation and atmospheric pollution monitoring [32].

7.5.3 *Photoacoustic Spectroscopy and Surface Acoustic Waves*

Photoacoustic spectroscopy (PAS) is a substance identification technique which relies on the illumination of a sample with specific wavelengths absorbed by the analyte. Part of the electromagnetic energy absorbed is transformed into heat, which translates into a physical expansion of the molecules and a subsequent pressure wave, which is measured with a microphone. The amplitude of this wave is proportional to the quantity of molecules that absorbed the initial electromagnetic energy, which also gives us the concentration.

In general, it is possible to obtain good measurement performance, even though selectivity is not high. Especially for ozone, sensitivity is not as high as photometric sensors. There are recent works using PAS to detect ozone through the application of visible light [33]. In literature, most works mention a minimum detection limit of 1–2 ppm [34–36], which narrows the scope of possible practical applications.

A recent technology that has shown promising results to measure gas concentrations in air, including ozone, is the use of surface acoustic waves (SAW). This is a type of sensor that benefited from the advances in nanotechnology. The physical principle of detection consists on the electro-mechanical coupling between resonating piezoelectric structures. The presence of an analyte changes the characteristics of this structure, such as mass or conductivity, also changing the propagation parameters of the acoustic wave [37]. It is common to represent sensitivity of this type of sensor with unit Hz/ppm when the resonant frequency is the parameter measured, or deg/ppm when phase is measured instead.

One of the biggest advantages of this type of sensor is the possibility of making passive RFID sensors. A sensor of this type with detection limit in the order of 60 ppb is reported in [38], although the objective was dosimetry and not instantaneous ozone concentration.

7.5.4 Solid State Sensors

Although electrochemical sensor are also *solid state*, this definition is usually applied to sensors that use thin films as sensitive elements, with thickness in the order of tens or hundreds of nanometers, often deposited through electron-beam physical vapor deposition (EB-PVD) [39].

This type of sensor has great potential, since thin films can be tuned in their physical structure or composition to favor gas detection for specific conditions and concentrations. Recent works mention capability of detection in the order of ppt (parts per trillion) [22, 40].

It is important to highlight that solid state sensors mentioned in literature are often impedimetric, where the relationship between the resistance of the sensitive element exposed to the measured gas and the resistance of the same element when exposed to air gives a sensitivity parameter.

Similarly to HMOS, solid state sensors usually need activation, also done with heat [39] or light [41]. There are thin film sensors being developed for room temperature operation, although not commercially available yet [42].

An interesting solid state sensor was produced in a small TO encapsulation, with a light source (LED) embedded in the sensitive structure itself. The authors claim to have measured ozone concentrations between 10 ppb and 200 ppm, which is an extremely high dynamic range, with low power requirements (50 mW) [43]. Unfortunately, this sensor does not seem to be commercially available.

7.5.5 Ozone Sensor Calibration

Even sensors with good technical features can be useless if not calibrated properly. Photometric sensors, in part due to their higher value and precision, usually come with a calibration certificate, which is valid for a given period of time.

Electrochemical sensors, including MOS, are much cheaper and, despite having good linearity, must be calibrated with a reference equipment—usually a photometric sensor. Few electrochemical sensor manufacturers provide factory calibration as an optional service and, when available, can cost more than the sensor itself.

The linearity of electrochemical sensors makes calibration fairly easy, since in theory you only need two data points to find the angular coefficient α:

$$C = \alpha V_{gas} + V_0, \tag{7.7}$$

where C in concentration of the measured gas, V_{gas} is the voltage at a given concentration and V_0 is the voltage at concentration zero. The use of a voltage signal assumes that the original signal has already been amplified, often through electronic circuits embedded in the same board as the sensor itself. However, this linearity is never ideal and this technique ignores the influence of temperature and humidity, for example. It does not take into consideration the cross-sensitivity with other gases as well. That is why more complex calibration methods have been developed.

A large study done in the European Union (MACPoll—Metrology of Chemical Pollutants in Air) evaluated many ozone sensors available commercially. The results are very interesting, because they show technical characteristics that are not supplied by the manufacturers, such as correlations between different variables, hysteresis, drift, etc [44]. Some of the most important conclusions of this study are:

- Amperometric and metal-oxide sensors are affected by changes in ambient temperature and humidity, but the change in behavior for amperometric sensors involves hysteresis, which is much difficult to compensate. In general, the influence of humidity changes (40%–80%) tends to have a larger impact than temperature changes (12–32°C).
- The cross-sensitivity of NO_2 in electrochemical ozone sensors is high, with a coefficient close to 1. Metal oxide sensors do not suffer as much from this problem.
- The long term stability (60–200 days) of electrochemical sensors was satisfactory, while metal oxide sensors showed significant drift in the same period of time.

To each commercial sensor evaluated, the researchers suggested a multivariate calibration model, taking into account the main factors that interfere with the measurements. That would allow for better estimates of the real ozone concentrations, assuming that the other variables are also being measured. For one of the sensors evaluated, the suggested calibration model is [45]:

$$O_3 = \frac{R_s - bNO_2 - cNO_2 \cdot H_2O - d}{a}. \tag{7.8}$$

The parameters a, b, c and d can be obtained from the linear regression of reference measurements taken during the calibration stage and R_s is the ozone sensor response.

A more sophisticated method for ozone sensor calibration consists in the use of artificial neural networks (ANNs). In [24], neural networks are compared to linear regression and multivariate linear regression. The best results are achieved by the neural networks, using O_3, NO_2 and CO sensors simultaneously. Curiously, humidity and temperature effects were also compensated by the neural networks, even though they were not being measured directly. This can be a result of the different way these parameters affect each of the three sensors.

7.6 Ozone Sensor Placement

To measure ozone concentration in a hydrogenerator, either a continuous monitoring system or a portable device can be used. Both methods have advantages and drawbacks. A continuous monitoring system will usually contain several sensors or air sample intakes spread through the machine. The position of each sensor should be as close as possible to the stator core, downwind from the cooling air direction and, ideally, in symmetrical locations. This can help locate points with more surface PDs.

If a portable equipment is used as means to obtain measurements periodically, there may be more restrictions on where to collect the samples, especially if the machine is in operation (as opposed to shut down, during off-line testing). It is recommended to make a number of measurements around the machine, considering the ventilation pattern, writing down the approximate placement of each one and keeping the records in order to evaluate trends over time. If the machine has water coolers, the closest accessible place during operation is usually behind those coolers [46].

Typical concentrations are in the level of dozens or hundreds of ppb (parts per billion), so when measuring ozone in a generator for the first time, it is convenient to have an equipment with good precision in this range. Portable devices usually rely on optical sensors, therefore with large dynamic range; if electrochemical sensors are used, then it may be necessary to establish a baseline measurement first, in order to specify the sensor with most adequate range.

Periodic ozone measurements can be done roughly at the same intervals as PD tests. There is no universal rule for this, but reasonable periods are every 6 months for a machine with low PD levels or every 3 months (or even more frequently) for situations where there is some concern about the insulation.

As mentioned before, ozone alone does not allow for a comprehensive diagnostic of the insulation's condition, but in generators equipped with PD sensors, for instance, the ozone concentrations can help identify the root cause of a failure mechanism.

7.7 Ozone Measurement in Current Electrical Machine Standards

The IEEE 1129 guide for the monitoring of synchronous machines above 10 MVA has one paragraph about ozone, mentioning that its presence is an indicative of partial discharges, while also accelerating the degradation of the insulating materials (chemical attack).

The IEEE 1434 guide for partial discharge measurements in AC machines also has a similar paragraph, adding that the sample source is critical, since the ozone concentration is affected by environmental conditions. It also mentions that ozone monitoring can be done in an offline machine, but the greatest benefit is measuring ozone with the machine in operation.

The IEEE 492 guide for operation and maintenance of hydrogenerators mentions briefly that the amount of ozone present inside the generator or around it can be an indicative of surface partial discharges.

No standard gives more objective information about how to actually measure ozone concentrations or how to compare it with other variables.

7.8 Other Gases

Ozone tends to be singled out as the only gas whose concentration can help identify partial discharges in a rotating machine, but more research is needed towards identifying other gases which may have the same potential. Air itself contains 78% of Nitrogen gas (N_2), so this molecule is also broken down through the same processes described before and compounds such as NO_x may also be an indicative of surface discharges. It is known that the concentration of NO_2 (Nitrogen dioxide) in air increases with the discharge voltage between electrodes [47]. Similarly to ozone, there are optical sensors in the market, which measure absorbance at 405 nm in order to determine the concentration of NO_x, and also electrochemical sensors, which are less expensive.

Although this chapter focused on the breakdown of air through electrical discharges, the stator insulation of modern machines is commonly made of mica and epoxy. Epoxy is a large class of resins with very diverse properties and molecular structures. Nevertheless, all of them are organic polymers and when subject to electrical discharges can be broken down into other components, some of them gaseous. Studies have found that epoxy resins used as insulation, when subject to electrical discharges, can generate by-products such as acetylene, acetone and carbon dioxide [48]. It is reasonable to expect, then, that these substances can also be worth monitoring and could add value from the perspective of predictive maintenance.

Recent works have shown that the type of insulation defect also influences the quantity of gases created by the discharges. In [49], four different types of insulation defects are compared and the concentration of O_3, NO and NO_3 measured. The authors suggest that the proportions between O_3/NO_x and NO_2/NO may be used to identify the types of faults, in a similar way that problems in power transformers are diagnosed through the analysis of dissolved gases in the insulating oil.

Acknowledgments This chapter was written as a part of the research project *P&D ANEEL PD-00642-2905/2020, "Investigation and methodology of ozone concentration analysis in hydrogenerators as a predictive maintenance tool"*. We would like to thank ANEEL (Brazilian National Agency of Electrical Energy) and the companies ENERCAN and BAESA.

References

1. Schoenbein, C. F. (1843). On the odour accompanying electricity, and on the probability of its dependence on the presence of a new substance; by C. F. Schœnbein, Professor of Chemistry, Bâle, communicated in a letter to Michael Faraday. *Abstracts of the Papers Printed in the Philosophical Transactions of the Royal Society of London*, *4*, 226–226.
2. Brasseur, G. P. (2020). The importance of fundamental science for society: The success story of ozone research. *Perspectives of Earth and Space Scientists*, *1*(1).
3. Chen, J., & Wang, P. (2005). Effect of relative humidity on electron distribution and ozone production by DC coronas in air. *IEEE Transactions on Plasma Science*, *33*(2), 808–812.
4. Wadhwa, C. L. (2006). *High voltage engineering* (2nd ed., pp. 10–14). New Age International.
5. Highfield, J. S. (1905). *The Electrician*, *54*, 573.
6. Epstein, J. (1907). Testing of electric machinery and of materials for its construction. *Journal of the Institution of Electrical Engineers*, *38*(181), 28–62.
7. Glazebrook, R., Skinner, C., Patchell, W., et al. (1907). Discussion on testing of electric machinery and of materials for its construction. *Journal of the Institution of Electrical Engineers*, *38*(181), 62–103
8. Fleming, A., & Johnson, R. (1911). Chemical action in the windings of high-voltage machines. *Journal of the Institution of Electrical Engineers*, *47*(209), 530–551.
9. Lowry, T. M. (1912). The oxidation of atmospheric nitrogen in presence of ozone. *Journal of the Chemical Society, Transactions*, *101*, 1152–1158.
10. Davis, W. Jr. (1972). Ozone formed by high-voltage transmission line coronas. Oak Ridge National Lab., Tenn. (USA), Technical Report, Central Files Report No. 72-7-25.
11. Lunt, R. (1959). The mechanism of ozone formation in electrical discharges. *Advances in Chemistry*, 286–303.
12. Pekárek, S., Rosenkranz, J., & Khun, J. (2004). Ozone generation from two hollow needles to plate electrical discharges enhanced by an airflow. In *12th international congress on plasma physics, Nice, France*.
13. Batakliev, T., Georgiev, V., Anachkov, M., Rakovsky, S., & Rakovsky, S. (2014). Ozone decomposition. *Interdisciplinary Toxicology*, *7*(2), 47–59.
14. Rodriguez, R., & Rux, L. (1997). Ozone abatement in air-cooled hydroelectric generators. U.S. Dept. of the Interior, Bureau of Reclamation, Washington DC, Technical Report.
15. Cestaro, G., & Fernandes, M. (2016). Fault detection in components of synchronous motors through online partial discharge measurements. In *IRMC Conference, Scottsdale, AZ.*
16. Stone, G. C., Culbert, I., Boulter, E. A., & Dhirani, H. (2014). *Electrical insulation for rotating machines: design, evaluation, aging, testing, and repair*. Wiley Blackwell.
17. Lépine, L. (2002). Mesure d'ozone dans les centrales hydroélectriques d'hydro-québec. Hydro-Québec, Montréal, Québec, Canada, Hydro-Québec Internal Report, IREQ-2002-069.
18. Binder, E., Draxler, A., Muhr, M., et al. (1999). Effects of air humidity and temperature to the activities of ex- ternal partial discharges of stator windings. In *1999 Eleventh international symposium on high voltage engineering, London, UK*.
19. Spec Sensor Operation Overview. (2016). https://www.spec-sensors.com/wp-content/uploads/2016/05/SPEC-Sensor-Operation-Overview.pdf. SPEC Sensors, LLC.
20. David, M., Ibrahim, M. H., Idrus, S. M., & Marcus, T.C.E. (2015). Fundamental review to ozone gas sensing using optical fibre sensors. *TELKOMNIKA (Telecommunication, Computing, Electronics and Control)*, *13*(4), 1133.
21. Simões, F., & Xavier, M. (2017). 6 - electrochemical sensors. In A. L. Da Róz, M. Ferreira, F. de Lima Leite, & O. N. Oliveira (Eds.). *Nanoscience and its applications*. Micro and Nano Technologies (pp. 155–178). William Andrew Publishing.
22. Srinives, S., Sarkar, T., Hernandez, R., & Mulchandani, A. (2017). Potassium iodide-functionalized polyaniline nanothin film chemiresistor for ultrasensitive ozone gas sensing. *Polymers*, *9*(12), 80.
23. Schenck, E. H., Burke, P. A., & Centanni, M. A. (2018). Patent us-10001455-b2.

24. Spinelle, L., Gerboles, M., Villani, M. G., Aleixandre, M., & Bonavitacola, F. (2015). Field calibration of a cluster of low-cost available sensors for air quality monitoring. Part A: Ozone and nitrogen dioxide. *Sensors and Actuators B: Chemical, 215*, 249–257.
25. Ripoll, A., Viana, M., Padrosa, M., et al. (2019). Testing the performance of sensors for ozone pollution monitoring in a citizen science approach. *Science of The Total Environment, 651*, 1166–1179.
26. Lewis, A., & Peltier, W. R., E. (2018). von Schneidemesser, Low-cost sensors for the measurement of atmospheric composition: Overview of topic and future applications. World Meteorological Organization (WMO), Geneva, Switzerland, Technical Report.
27. Gorshelev, V., Serdyuchenko, A., Weber, M., Chehade, W., & Burrows, J. (2014). High spectral resolution ozone absorption cross-sections part 1: Measurements, data analysis and comparison with previous measurements around 293 k. *Atmospheric Measurement Techniques*, *7*(2), 609–624.
28. Mayerhöfer, T. G., Pahlow, S., & Popp, J. (2020). The bouguer-beer-lambert law: Shining light on the obscure. *Physical Chemistry, Chemical Physics*, *21*(18), 2029–2046.
29. Andersen, P. C., Williford, C. J., & Birks, J. W. (2010). Miniature personal ozone monitor based on UV ab- sorbance†. *Analytical Chemistry*, *82*(19), 7924–7928.
30. Grzg Andritz. (2018). Andritz cooling air analysis - dia tech caa-o3 product brochure. HP.DIATECH_CAA-03 01.e.04.12. Andritz Hy- dro, Vienna, Austria. https://grzg.andritz.com/c/com2011/00/02/51/25190/1/1/0/228048199/hy-diatech-caa-03.pdf.
31. De Maria, L., & Rizzi, G. (2009). Ozone sensor for application in medium voltage switchboard. *Journal of Sensor*, *2009*, 1–5.
32. Ray, J. D., Stedman, D. H., & Wendel, G. J. (1986). Fast chemiluminescent method for measurement of ambient ozone. *Analytical Chemistry*, *58*(3), 598–600.
33. Keeratirawee, K., & Hauser, P. C. (2021). Photoacoustic detection of ozone with a red laser diode. *Talanta*, *223*, 121890
34. Köhring, M., Willer, U., Böttger, S., Pohlkötter, A., & Schade, W. (2012). Fiber-coupled ozone sensor based on tuning fork-enhanced interferometric photoacoustic spectroscopy. *IEEE Journal of Selected Topics in Quantum Electronics*, *18*(5), 1566–1572.
35. Böttger, S., Köhring, M., Willer, U., & Schade, W. (2013). Off-beam quartz-enhanced photoacoustic spectroscopy with LEDs. *Applied Physics B*, *113*(2), 227–232 .
36. Chien, H.-T., Wang, K., Sheen, S.-H., & Raptis, A. C., Standoff detection of ozone in an open environment using photoacoustic spectroscopy technique. *Applied Physics Letters*, *100*(10), 104102.
37. Devkota, J., Ohodnicki, P., & Greve, D., SAW sensors for chemical vapors and gases. *Sensors*, *17*(4), 801.
38. Westafer, R. S., Levitin, G., Hess, D. W., Bergin, M. H., & Hunt, W. D. (2014). Detection of ppb ozone using a dispersive surface acoustic wave reflective delay line with integrated reference signal. *Sensors and Actuators B: Chemical*, *192*, 406–413.
39. Mastelaro, V. R., Zılio, S. C., da Silva, L.F. et al. (2013). Ozone gas sensor based on nanocrystalline SrTi1-xFexO3 thin films. *Sensors and Actuators B: Chemical*, *181*, 919–924.
40. B. de Lima, Komorizono, A., Silva, W., et al. (2021). Ozone detection in the ppt-level with rGO-ZnO based sensor. *Sensors and Actuators B: Chemical*, *338*, 129779.
41. Wu, C.-H., Jiang, G.-J., Chang, K.-W., et al. (2018). Analysis of the sensing properties of a highly stable and reproducible ozone gas sensor based on amorphous in-ga-zn-o thin film. *Sensors*, *18*(2), 163.
42. Sui, Y., Liang, H., Chen, Q., Huo, W., Du, X., & Mei, Z. (2020). Room-temperature ozone sensing capability of IGZO-decorated amorphous ga2o3 films. *ACS Applied Materials & Interfaces*, **12**(7), 8929–8934.
43. Wang, C., Becker, R., Passow, T., et al. (2011). Photon stimulated sensor based on indium oxide nanoparticles i: Wide-concentration-range ozone monitoring in air. *Sensors and Actuators B: Chemical*, *152*(2), 235–240.
44. European Commission. (2013). Protocol of evaluation and calibration of low-cost gas sensors for the monitoring of air pollution - joint research centre - institute for environment and sustainability. European Union, Technical Report.

45. European Commission. (2014). Laboratory and in-situ validation of ozone microsensors, αsense, model b4 o3 sensors. task 4.3, testing protocol, procedures and testing of performances of sensors. European Union, Technical Report.
46. Lépine, L. (2007).Understanding ozone distribution inside stator core and measurements inside air-cooled gener- ators to assess partial discharges problems. In *Iris rotating machine conference, San Antonio, TX.*
47. Pekárek, S. (2017). Experimental study of nitrogen oxides and ozone generation by corona-like dielectric barrier discharge with airflow in a magnetic field. *Plasma Chemistry and Plasma Processing*, *37*(5), 1313–1330.
48. Zhang, X., Wu, Y., Chen, X., Wen, H., & Xiao, S. (2017). Theoretical study on decomposition mechanism of insulating epoxy resin cured by anhydride. *Polymers*, *9*(12), 341.
49. Li, K., Javed, H., Zhang, G., & Plesca, A. T., Analysis of air decomposition by-products under four kinds of partial discharge defects. *IEEE Transactions on Dielectrics and Electrical Insulation*, *24*(6), 3713–3721.

Appendix A
International Standards and Norms Related to Partial Discharges

Partial discharge monitoring in isolation systems of synchronous machines still is the main diagnostic technique in use with proven efficacy [1]. According to the International Electrotechnical Commission (IEC) as it is defined in IEC 60270 [2], partial discharges are localized electrical discharges that partially short-circuit the insulation between conductors, and which may or may not occur adjacently to conductors. The main standards governing off-line and on-line PD measurements are respectively IEC 60034-27 [3] and IEC 60034-27-2 [4].

In order to carry out PD measurements, it is required a reliable measuring circuit because capacitive couplers are connected directly to the generator output bus, i.e., at high voltage terminals. This way, capacitive couplers must meet a series of requirements. The couplers are made of high quality virgin mica sheets. The epoxy material used in the coupling fabrication should be flame-retardant with excellent stability at high temperatures. The properties of electrical tracking damage must be tested according to ANSI/IEEE C37.20.2 [5]. Independent endurance tests (IEEE Std 1043 [6]) should show that the couplers support more than 1000 hours at 30 kVrms. The couplers must also be tested against the parameter B.I.L. (Basic Impulse Level)[7], capable of withstanding, without rupture events, five 150 kV pulsed tests (95 kV required by ANSI C37.20.2 [8]) and application of constant voltage level of 150 kV (50 kV are required by ANSI C37.20.2 [8]).

By means of a measuring device consisting of a signal processing unit performing phase synchronization and producing visualization outputs, it is possible to interpret the PD signals from the stator winding. PD signals are a symptom of a fault process that can evolve into failure. It is usually not possible to specify exactly a level of a PD magnitude where there is a high risk of insulation failure. However, meaningful interpretation of results obtained online is possible using various some sources of information, such as machine design knowledge base, maintenance history, visual inspections, operational conditions besides various off-line tests that are also used for quality assurance (IEEE Std 1799 [9]).

V. Dmitriev et al., *Partial Discharges in Hydroelectric Generators*, Power Systems,
https://doi.org/10.1007/978-3-031-36604-8

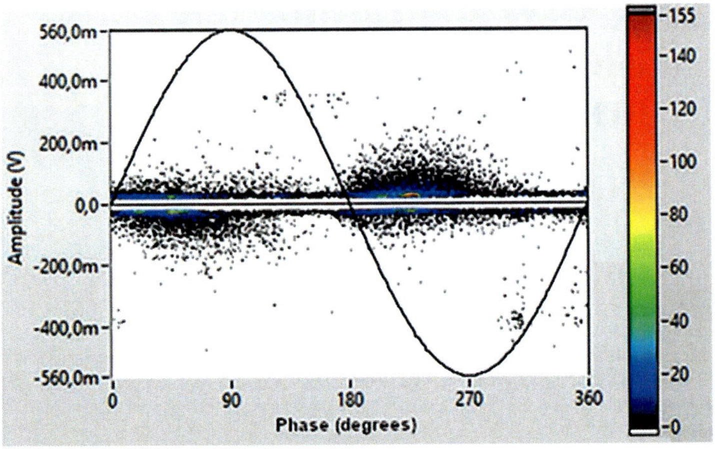

Fig. A.1 Example of a PD φ-q-n map, color-coded for mapping PD pulse count

In general, interpretation of online PD results should be performed in two steps. First, it is critical for any maintenance planning to know if there are isolation problems, as indicated by significant PD activity. If PD activity is confirmed, it needs a more detailed assessment. Since the degree of deterioration, and hence the risk of insulation failure, depends greatly on the specific type of PD, it is crucial to have good information about the source of any significant PD activity, such as the type and the possible location within the machine stator winding.

A powerful means of interpreting online PD data is evaluation of the tendency of specific PD parameters over time. For new machines, it is essential for a reliable trend assessment to have an initial record of PD activity to be used for comparison with subsequent regular or continuous PD measurement results. For old machines, the reference parameters will be the first measurements taken. In this sense, IEC 60034-27-2 [4] describes various discharge patterns present in rotary machines. To evaluate the condition of the stator winding, *Phase Resolved Partial Discharge* map (PRPD), also referred to as φ-q-n map (Fig. A.1), recorded during periodic measurements or during continuous monitoring, should be used to determine the specific types of PD activities in machine stator winding [3, 4].

By classifying the φ-q-n patterns, one can separate several PD sources from each other and observe their behavior separately. When the specific type and location of any PD activity within the winding is known, it is possible also to evaluate the risk associated with the PD sources. Since each PD process can have its own critical level of PD magnitude, it is not recommended to use PD magnitude alone as a risk indication of premature failure. The norm IEC 60270 [2] describes the high-voltage test techniques for partial discharge signal measurements in, for instance, cables, rotating machines or gas insulated switchgears. The standard IEEE P1434/D1.1 [10, 11] is specifically written for defining standard procedures for measuring partial discharges in AC electric machinery. IEEE P1434/D1.1 discusses both on-line and off-line PD measurement procedures for windings of any type as well as measurements on individual coils and bars. IEC 60034-27 [3] specifies off-line partial discharge measurement procedures of stator winding insulation of rotating electrical machines and IEC 60034-27-2 [4] is concerned with on-line PD measurement methods. Finally, IEEE Std 1799 [9] recommends practices for quality control testing of external discharges on stator coils, bars, and windings.

Appendix B
Antenna Definition and Basic Parameters

B.1 Antennas

According to the 2014 international standard IEEE 145 [12], which defines the engineering parameters for Antennas, an antenna is defined as a device capable of receiving or radiating radio waves. Similarly, the Brazilian telecommunication national agency (in Portuguese: *Agência Nacional de Telecomunicações - ANATEL*), defines antenna as a device for, in any telecommunication system, radiating or capturing electromagnetic waves [13]. Antennas can include circuits, which can be incorporated into the device so that they contribute, in a smart way or not, to define its radiating characteristics.

In a transmission system, there are basically three elements: the Antennas (receiving and transmitting electromagnetic waves) and the environment in which the waves propagate [14], producing the propagation channel. In RF communication systems, the transmitting system is usually the one that needs more energy in order to operate properly. This means that its design is often more critical. The signal emitted by the transmitting antenna must have sufficient power and must be well orientated so that the receiving antenna is capable of capturing the signal, establishing communication [15]. However, when antennas are used to perform partial discharges radiated signal measurement, one must study/optimize its receiving parameters (or transmitting parameters, which are coincident due to the reciprocity principle) [15].

In order to the receiving antenna gets maximum power from PD electromagnetic waves, it is needed that the maximum of the radiation pattern is directed the PD emission. If that is the case, the received electromagnetic wave produces maximum voltage in antenna terminals [15]. For knowing the behavior of an antenna, its radiation pattern, radiated power, radiation intensity, directivity, gain, bandwidth, beamwidth, efficiency, polarization, input impedance and S_{11} (or return loss) must be obtained.

V. Dmitriev et al., *Partial Discharges in Hydroelectric Generators*, Power Systems,
https://doi.org/10.1007/978-3-031-36604-8

Classification of Antennas

According to [14], the antennas can be classified according to its physical structure:

(a) Filamentary antennas: there are various shapes for filamentary antennas, which are made out of filament(s) of wire, such as the straight wire dipole antenna, loop antennas (mainly in circular or square geometries) and the helical antenna;
(b) Aperture antennas, which can be the conical or pyramidal horns or waveguides;
(c) Microstrip antennas: its simplest form consist of a metallic transmission line put over a grounded substrate;
(d) Antenna arrays: are composed of various coupled antennas. Various types of antennas can form arrays. For instance we have dipole arrays, Yagi-Uda antennas or the log-periodic antennas;
(e) Antennas with reflectors: classic examples are the parabolic antenna, which are used to long-range applications.
(f) Lens antennas: contain transparent dielectric parts used to refract and focus electromagnetic waves to be transmitted or received in parallel rays (similarly to what glass lenses do to light), thus reducing spatial power spreading to undesired directions.

B.2 Fundamental Parameters of Antennas

B.2.1 Radiation Pattern

The antenna radiation pattern is defined as a mathematical function or a graphical representation of the antenna radiation strength (usually gain) as a function of spatial coordinates [14]. The radiation patterns can be classified in isotropic, directional and omnidirectional.

An isotropic radiator is defined as a hypothetical antenna, with no losses, that has a given single radiation strength in all directions (thus, its radiation pattern would be a perfect sphere, as it can be seen in Fig. B.1a). An omnidirectional antenna is defined a radiator that has a radiation pattern essentially uniform and, thus, non-directional, on a given plane (i.e., a circle on that plane), usually with nulls established orthogonally to that non-directional radiating plane. An omnidirectional pattern has a geometry similar to that of a donut (see Fig. B.1b). Finally, a directional antenna has the capacity of radiating or receiving waves more efficiently to/from specific directions, usually as designed by antenna engineers. It is illustrated by Fig. B.1c.

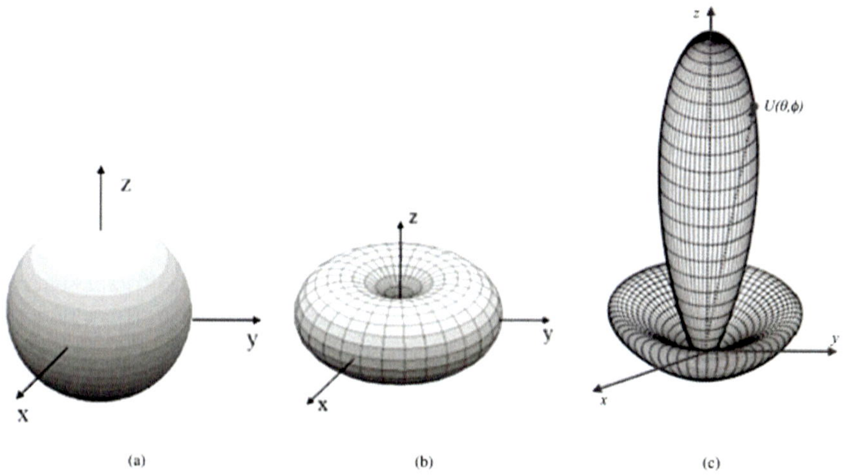

Fig. B.1 Illustrative examples of radiation patterns: (**a**) isotropic; (**b**) omnidirectional; (**c**) directional. Source: adapted from [14]

B.2.2 Radiation Power Density

A vector quantity used to describe the power density associated to the electromagnetic waves and its propagation direction is the instantaneous Poynting vector $\overrightarrow{W}(W/m^2)$. It is defined as the cross product between the instantaneous electric field $\overrightarrow{E}$ (V/m) and the instantaneous magnetic field $\overrightarrow{H}$ (A/m). Mathematically, it can be written as

$$\overrightarrow{W} = \overrightarrow{E} \times \overrightarrow{H}. \tag{B.1}$$

The instantaneous radiated power P_{rad} can be obtained by integrating (B.1) over a closed surface S defined mathematically around the radiator. Mathematically, one has

$$P_{rad} = \oint_S \overrightarrow{W}_{rad} \cdot d\overrightarrow{s}. \tag{B.2}$$

If an averaged radiated power density $\overrightarrow{W}_{\text{av}}$, at a given frequency, must be calculated, one must compute the Fourier transforms of $\overrightarrow{E}$ and $\overrightarrow{H}$, which are represented here by $\overrightarrow{e}$ and $\overrightarrow{h}$ and finally calculate

$$\overrightarrow{W}_{\text{av}} = \frac{1}{2}\Re[\overrightarrow{e} \times \overrightarrow{h}]. \tag{B.3}$$

Thus, averaged radiated power $\bar{P}_{\text{av}}$ at the considered frequency is given by

$$\bar{P}_{\text{av}} = \oint_S \vec{W}_{\text{av}} \cdot d\vec{s}\,. \tag{B.4}$$

B.2.3 Regions of Far and Near Fields

According to [14], the space that involves an antenna is usually subdivided into three regions: (a) region of reactive (non-radiative) near-field, (b) region of radiative near-field (Fresnel region) and (c) region of far-field (Fraunhofer region), as it is illustrated in Fig. B.2. The boundaries limiting the three region are not uniquely defined, and diverse criteria were established. However, in practice, the measurements of RF signals are performed in the far field region, so that all radiation characteristics are properly defined [15].

The region of the reactive near-field is defined as the "portion of the near-field region immediately around the antenna, where the reactive field predominates" [14]. To the majority of the antennas, the external frontier of this region is generally established at the distance $R = 0.62\sqrt{D^3/\lambda}$ measured from the source point, where λ is the wavelength and D is the larger dimension of the antenna. To a very short dipole or equivalent radiating element, that boundary is commonly established at the distance of $\lambda/(2\pi)$ measured from the antenna external surface.

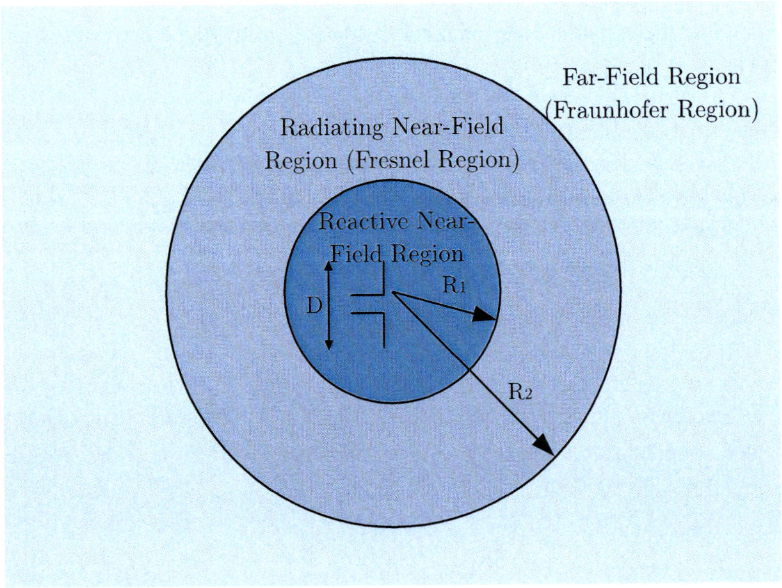

Fig. B.2 Field regions of an antenna. Source: adapted of [14]

The radiative near-field region (Fresnel region) is defined as the antenna field region between the reactive near-field region and the far-field volume, in which predominates the radiative field characterized by the fact that the angular distribution of the fields depends on the distance measured from the antenna. The boundaries of this region are defined by the distances $R = 0.62\sqrt{D^3/\lambda}$ and $R = 2D^2/\lambda$, both measured from the antenna.

The far-field region (Fraunhofer region) is defined as the field region of an antenna where the angular distribution of the fields does not depend on the distance to the antenna. The Fraunhofer region is usually established at distances greater than $2D^2/\lambda$ of the antenna. The radiation patterns of antennas are defined in the far-field region.

B.2.4 Radiation Intensity

The radiation intensity U is defined as "the power radiated by the antenna by unit of solid angle" [14]. The radiation intensity is a far field parameter and it can be obtained by multiplying the radiation density by r^2 (the squared distance between the source and the measuring point). It is given by

$$U = r^2 W_{rad}. \tag{B.5}$$

B.2.5 Directivity

The directivity of an antenna is defined as the ratio between the radiation intensity $U(\theta, \phi)$ in given angular spherical coordinates and the average radiation intensity U_0, which is the radiation intensity of an isotropic source. If angle coordinates are not specified, one usually takes $\max\{U(\theta, \phi)\}$. It can be understood as the ability of the antenna in concentrate the radiated power over space [15]. Mathematically, we have

$$D = \frac{U}{U_0} = \frac{4\pi U}{P_0}, \tag{B.6}$$

where D is directivity (dimensionless), U is radiation intensity (W/unit of solid angle); U_0 is the radiation intensity of an isotropic source (W/unit of solid angle) and P_{rad} is the total radiated power (W) by the antenna.

B.2.6 Antenna Polarization

The polarization of an antenna in a given far-field point of space is defined as the polarization of the transmitted wave (radiated) by the antenna. In practice, the polarization of the radiated wave is defined by the direction of electric field.

Antenna polarization is function of the spherical angles. However, polarization can be considered to be the direction of electric field at spherical coordinates where maximum gain is obtained.

B.2.7 Return Loss

In simple terms, the antenna return loss R_L is a measure of the power fraction returning to the transmitter which is connected to the transmitting antenna. This loss (power return) is associated to impedance mismatch between the antenna complex impedance Z and the transmitter internal impedance Z_0. Antenna impedance Z is given by $Z = \tilde{V}/\tilde{I}$, in which $\tilde{V}$ and $\tilde{I}$ are the Fourier transforms of voltage and current obtained across the antenna's terminals. Thus, Z is a function of frequency.

Complex reflection coefficient Γ between Z and Z_0 is given by

$$\Gamma = \frac{Z - Z_0}{Z + Z_0}. \tag{B.7}$$

Finally, return loss, as a function of frequency, is defined by

$$R_L(\text{dB}) = -20 \log_{10} |\Gamma| . \tag{B.8}$$

In most cases, antenna operation band(s) is(are) defined as the frequency band(s) in which $R_L \leq -10\,\text{dB}$. Notice that this parameter can be obtained when the antenna is connected to a receiver. In that case, Z_0 is the receiver internal impedance.

Antenna bandwidth Δf is the difference between the upper cut-off frequency f_2 and the lower cut-off frequency f_1. It is given by

$$\Delta f = f_2 - f_1. \tag{B.9}$$

Frequently, f_1 and f_2 are respectively defined as the lower and higher frequencies at which R_L crosses the level of $-10\,\text{dB}$.

Appendix C
Partial Discharges Numerical Modelling: Overview of the Finite-Difference Time-Domain (FDTD) Method

C.1 Introduction

In 1965, Frank Harlow published a work which demonstrated a method for obtaining numerical solutions of the Navier-Stokes equation [16] based on finite differences in time domain, for incompressible fluid flow with a free surface in two dimensions. The following year, Kane Yee publishes a similar work [17] that introduces a technique capable of directly and relatively simply and elegantly solving the Maxwell equations numerically in time domain. This method, known as the Yee Algorithm, is based on two main aspects: (1) a discrete geometric distribution of the electric field $\overrightarrow{E}$ and magnetic field $\overrightarrow{H}$ components in cells, in order to satisfy both Faraday's Law and Ampere's Law in differential and integral forms and (2) approximating spatial and temporal derivatives using algebraic differences (finite differences), in order to obtain explicit equations for calculating all field components over space-time.

In addition, the components of the electric field are always displaced in time in "half" a time step from magnetic field components, so as to satisfy central temporal derivatives along with the aforementioned spatial distribution of components (in addition to avoiding problems related to matrix inversions). This way, a quite robust method, capable of generating full-wave solutions, arises.

For some time, the method was not widely applied due to the fact that it requires considerable amounts of computational resources, which was quite restricted and limited mainly in the first decade of the method's existence. However, during the 1970s, Allen Taflove and Brodwin applied Yee's technique to solve three-dimensional problems related to the electromagnetic interaction with material media [18], in addition to modeling bio-electromagnetic problems in a pioneering way [19]. In the same year, Taflove also publishes the stability condition for the Yee method [18]. In 1980, Taflove validated numerical results regarding metal cavities and disseminates the term FDTD (*Finite-Difference Time-Domain Method*) [20]. Since then, the application of the technique for solving a large number of problem

V. Dmitriev et al., *Partial Discharges in Hydroelectric Generators*, Power Systems,
https://doi.org/10.1007/978-3-031-36604-8

classes has produced a number of works that increase exponentially over the years [21].

At the time, as there were no techniques capable of truncating computational domains for properly modelling open problems, many truncation formulations were developed, such as, for example, those based on the Bayliss-Turkel operators [22], Mür's first and second order method [23], Higdon's technique [24] and Liao's technique [25]. These techniques are known as ABCs (Absorbing Boundary Conditions), and their main goal is to absorb waves impinging at the ends of the analysis regions, simulating this way wave propagation to infinity (thus enabling this kind of simulation in finite Yee grids). The most recent and most efficient ABCs are, to this day, those based on the idea of layers perfectly matched with the analysis region (PML), originally developed and implemented by Berenger [26]. In this work, the UPML (Uniaxial Perfectly Matched Layers) truncation technique [27] has been applied to truncate three-dimensional FDTD domains.

During the 1980s and 1990s, the FDTD method was expanded to local coordinate systems, which do not coincide with the Cartesian system, based on the work of Holland [28] and Mei [29] and later that of Lee [30]. The technique became known as FDTD in general coordinates or, more specifically, local non-orthogonal FDTD (LN-FDTD).

Other important techniques are those for modelling metallic cylindrical rods with diameters smaller than the spatial increments used. The technique known as Thin Wire [31, 32] was applied in this work to model the rods in such way to avoid high levels of space discretization.

The FDTD method and its variations have been applied to solve various types of problems, from antennas [33] to the design and analysis of high-frequency circuits (such as microprocessors) [34], radars [35], photonics [36], communication systems [37], periodic structures [38], grounding systems [39], partial discharges [40], graphene analysis [41], medicine [42], among others. The method is able to analyze the influence of various types of materials on the propagation of electromagnetic waves, such as homogeneous, non-homogeneous, anisotropic and dispersive [43]. Thus, a wide range of applications is observed in the literature for FDTD.

This appendix will attempt to show, in a very objective way, the theoretical basis necessary for understanding the FDTD method (Yee algorithm) in rectangular coordinates, the application of absorbing boundary conditions and excitation technique for modeling partial discharges.

C.2 The Yee Algorithm (FDTD Method) and Computer Implementation

Time domain Maxwell's equations are given by

$$\mu \frac{\partial \overrightarrow{H}}{\partial t} = -\nabla \times \overrightarrow{E} \tag{C.1}$$

and

$$\epsilon \frac{\partial \overrightarrow{E}}{\partial t} = \nabla \times \overrightarrow{H} - \overrightarrow{J}, \tag{C.2}$$

in which $\overrightarrow{E}$ is the electric field (V/m), $\overrightarrow{H}$ is the magnetic field (A/m), ϵ and μ are, respectively, the electric permittivity (farad/m) and magnetic permeability (henry/m). The term $\overrightarrow{J} = \sigma \overrightarrow{E}$ is the electric current density (A/m^2), in which σ is the electrical conductivity of the medium (S/m).

Equations (C.1) and (C.2), when expanded in rectangular coordinates, generate the scalar equations

$$\frac{\partial H_x}{\partial t} = \frac{1}{\mu}\left(\frac{\partial E_y}{\partial z} - \frac{\partial E_z}{\partial y}\right), \tag{C.3}$$

$$\frac{\partial H_y}{\partial t} = \frac{1}{\mu}\left(\frac{\partial E_z}{\partial x} - \frac{\partial E_x}{\partial z}\right), \tag{C.4}$$

$$\frac{\partial H_z}{\partial t} = \frac{1}{\mu}\left(\frac{\partial E_x}{\partial y} - \frac{\partial E_y}{\partial x}\right), \tag{C.5}$$

and

$$\frac{\partial E_x}{\partial t} = \frac{1}{\epsilon}\left(\frac{\partial H_z}{\partial y} - \frac{\partial H_y}{\partial z} - \sigma E_x\right), \tag{C.6}$$

$$\frac{\partial E_y}{\partial t} = \frac{1}{\epsilon}\left(\frac{\partial H_x}{\partial z} - \frac{\partial H_z}{\partial x} - \sigma E_y\right), \tag{C.7}$$

$$\frac{\partial E_z}{\partial t} = \frac{1}{\epsilon}\left(\frac{\partial H_y}{\partial x} - \frac{\partial H_x}{\partial y} - \sigma E_z\right), \tag{C.8}$$

where E_x, E_y, E_z and H_x, H_y, H_z are the scalar components of the electric field $\overrightarrow{E}$ and magnetic field $\overrightarrow{H}$, respectively. These components are, in general, functions of time t and the three Cartesian coordinates x, y, and z.

Equation (C.1), Faraday's law, mathematically shows how temporal variation of $\overrightarrow{B} = \mu \overrightarrow{H}$ (magnetic flux density vector) produces components of electric field circulating around $\overrightarrow{B}$. Ampére's law, similarly, shows that when there is time variation of $\overrightarrow{D} = \epsilon \overrightarrow{E}$, circulation of the magnetic field arises around $\overrightarrow{D}$. These two observations about Eqs. (C.1) and (C.2) are the pillars of the entire theory of propagation of electromagnetic waves in the macroscopic universe [44], on which Yee based the definition of his spatial and temporal distribution of the field

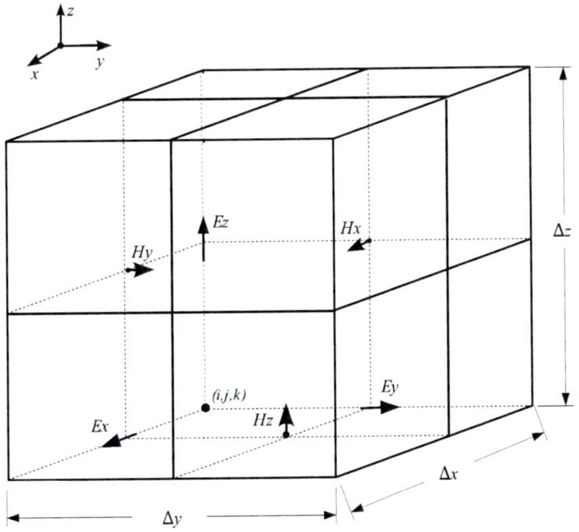

Fig. C.1 Spatial distribution of scalar components of $\overrightarrow{E}$ and $\overrightarrow{H}$ in Yee cell. This cell is indexed by (i, j, k)

components of his numerical algorithm. The spatial placement of these components is illustrated in Fig. C.1: the orthogonal Yee cell.

When propagating wave simulations are performed in a given structure, the cell shown in Fig. C.1 is applied throughout the region of interest (analysis region) to show the components of the fields $\overrightarrow{E}$ and $\overrightarrow{H}$. Each point of the mesh thus formed, as well as the components shown in Fig. C.1, is referenced by the discrete indices i, j, k. This means that a given position x, y, z (in meters) is addressed in the discrete space by i, j, k (corresponding cell number), so that $x = i\Delta_x$, $y = j\Delta_y$ and $z = k\Delta_z$, where Δ_x, Δ_y and Δ_z are the dimensions of the Yee cells.

The spatial distribution shown in Fig. C.1 is a discretized geometric representation of the equations in (C.1) and (C.2). It is possible to observe, using the right-hand rule, that a positive time variation of the E_z component (time derivative greater than zero) generates a magnetic field circulation around $E_z\hat{a}_z$, so that, to update this component (Eq. (C.8)), it is necessary to take into account the components of the magnetic field $H_x(i, j - 1/2, k + 1/2)$ and $H_x(i, j + 1/2, k + 1/2)$ (characterizing the derivative of H_x with respect to y) and the components $H_y(i + 1/2, j, k + 1/2)$ and $H_y(i - 1/2, j, k + 1/2)$ (characterizing the derivative of H_y with respect to x). The same can be observed for the calculation of the $\overrightarrow{H}$ field components.

Figure C.1 shows that the electric field components are always centered with respect to four magnetic field components and vice versa. Obviously, this distance between the $\overrightarrow{E}$ and $\overrightarrow{H}$ field components (half the edge of the Yee cell) also results in a temporal interleaving between them (the sampling interval is Δ_t). This interleaving in time is commonly called *leapfrog*, and works as follows: after updating all the electric field components at a time $t = t_1$, the magnetic field component update is performed, based on the $\overrightarrow{E}$ components that have just been updated, for the time $t = t_1 + \frac{\Delta_t}{2}$. The subsequent update of the electric field

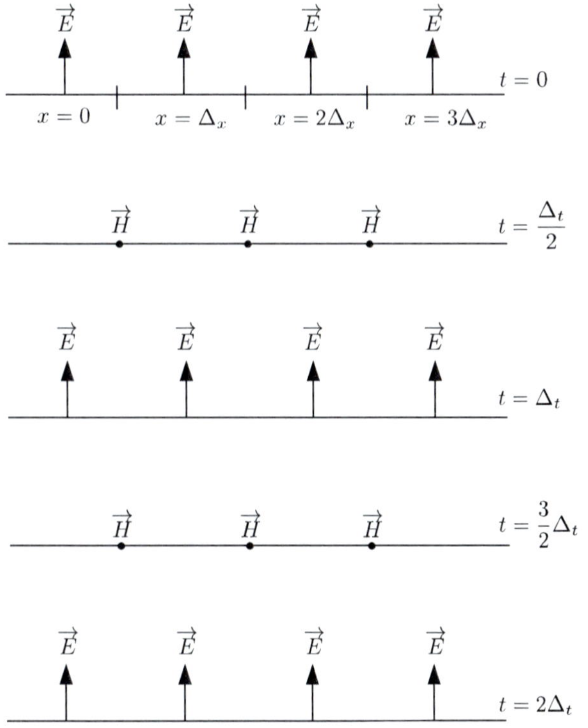

Fig. C.2 Leapfrog: time and space distributions of $\vec{E}$ and $\vec{H}$ (one-dimensional space)

components is performed for the time $t = t_1 + \Delta_t$ and is based on the magnetic field components at $t = t_1 + \frac{\Delta_t}{2}$, previously stored in memory. The process is repeated until the time interval necessary to complete the simulation, previously established, is completed. The process is illustrated in Fig. C.2 for the one-dimensional case. In Fig. C.2, the distribution of the $\vec{E}$ and $\vec{H}$ components is shown for a one-dimensional case of the Yee algorithm, interpolated in time and space.

In order to obtain the FDTD equations for $\vec{E}$ and $\vec{H}$, the following procedure is followed. Based on the considerations made previously about the FDTD method, it can be easily seen that the derivatives that appear in Eqs. (C.3)–(C.8) can be approximated by central derivatives, both with respect to time and space (see Figs. C.1 and C.2). The approximation

$$\frac{\partial f(x)}{\partial x} \approx \frac{f(x + \frac{1}{2}\Delta_x) - f(x - \frac{1}{2}\Delta_x)}{\Delta_x} \tag{C.9}$$

obtained from the second order truncation of the Taylor series, defines the concept of one-dimensional central derivative. The derivative at point x of the function f with respect to x is obtained from the value of the function at positions $x + \frac{1}{2}\Delta_x$ and $x - \frac{1}{2}\Delta_x$. Theoretically, the smaller the value of Δ_x, the better the approximation of the derivative. Computationally, this value is limited by the accuracy of the machine or the compiler.

One can easily test (C.9) by using known functions and values for x and Δ_x. For example, the derivative of $\log(x)$ (natural logarithm), is exactly $1/x$. Numerically, it can be seen that, if $f(x) = \log(x)$, $x = 2$ and $\Delta_x = 0.02$, $df(x)/dx \approx (\log(2 + 0.01) - \log(2 - 0.01))/0.02 \approx 0.500004166729162$, which approaches to 1/2 as Δ_x is reduced. Illustratively, if $f(x) = \sin(x)$, $x = 1$ and $\Delta_x = 0.01$, we have $df(x)/dx \approx (\sin(1 + 0.005) - \sin(1 - 0.005))/0.01 \approx 0.540300054611331$, which is fairly close to exact solution $\cos(1) = 0.540302305868140\cdots$. Therefore, (C.9) can be seen as a general formula to obtain first derivatives numerically, including those of (so far unknown) functions satisfying differential equations.

A function f depending on the spatial coordinates (x, y, z) and the time t is approximated by a sampled function f_s, i.e.,

$$f(t, x, y, z) \approx f_s^n(i, j, k) \tag{C.10}$$

where i, j and k are the indices of the spatial increments and n the temporal index. Thus, using index notations, (C.9) can be rewritten, for one-dimensional space, as

$$\frac{\partial f(x)}{\partial x} \approx \frac{f(i + 1/2) - f(i - 1/2)}{\Delta_x}, \tag{C.11}$$

where $x = i\Delta_x$, as previously stated.

For sake of didactic explanation, (C.4) and (C.8) are used here to illustrate computer implementation of the FDTD algorithm for one-dimensional wave propagating along x over time t. For this goal, we should compute H_y and E_z since electromagnetic propagation is characterized by transverse waves [43, 44], i.e., electromagnetic oscillations are perpendicular to the propagation direction (see Fig. C.3). Notice that H_z and E_y could have been chosen.

For representing one-dimensional propagation along x, notice that partial derivatives with respect to y and z in (C.4) and (C.8) must be disregarded. Furthermore, we consider non-conductive media in this example ($\sigma = 0$). Thus, we obtain, from (C.4) and (C.8), respectively,

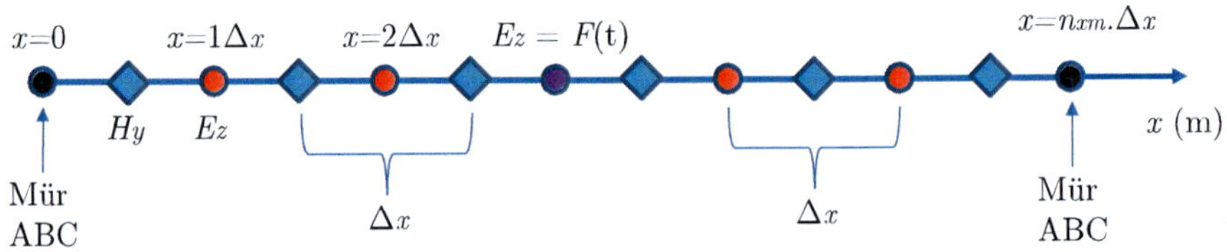

Fig. C.3 Example of FDTD-1D computational mesh for numerical modelling of wave propagation along x using E_z and H_y. Wave is excited using the function $F(t)$

$$\frac{\partial H_y}{\partial t} = \frac{1}{\mu}\left(\frac{\partial E_z}{\partial x}\right) \tag{C.12}$$

and

$$\frac{\partial E_z}{\partial t} = \frac{1}{\epsilon}\left(\frac{\partial H_y}{\partial x}\right). \tag{C.13}$$

By employing (C.11) to space and time derivatives in (C.12) and (C.13), and by rearranging terms, we see, using notations defined in (C.10) and the mesh shown in Fig. C.3, that

$$H_y^{n+1/2}(i+1/2) = H_y^{n-1/2}(i+1/2) + \frac{\Delta_t}{\mu}\left(\frac{E_z^n(i+1) - E_z^n(i)}{\Delta_x}\right) \tag{C.14}$$

and

$$E_z^{n+1}(i) = E_z^n(i) + \frac{\Delta_t}{\epsilon}\left(\frac{H_y^{n+1/2}(i+1/2) - H_y^{n+1/2}(i-1/2)}{\Delta_x}\right). \tag{C.15}$$

Real computer implementation of (C.14) and (C.15) is shown by Figs. C.4 and C.5 using the programming language C. Initialization of basic parameters and variables (such as universal constants, FDTD parameters, media parameters and fields) are seen in Fig. C.4. Main implementation of FDTD-1D algorithm is given in Fig. C.5.

Equations (C.14) and (C.15) are seen in lines numbered as 100 and 114 in Fig. C.5. Notice that both equations are implemented inside particular space loops (one space loop for each equation) controlled by using space index i. Furthermore, all FDTD space loops used to update E_z and H_y over the entire space, are implemented inside the time loop controlled by using time index n (see lines 92–120 in Fig. C.5). The implementation must be performed in the just described way because, as Maxwell's equations and classical physics dictate [44], advance in time must occur uniformly over the entire space. As an approximation resulting from central approximation for time derivatives (such as (C.9) for space), E_z and H_y must be calculated in an alternating arrangement regarding time and space (leapfrog sequence [43]). This feature is indicated by the "half" indexes in (C.14) and (C.14) and it is illustrated by Fig. C.2. "Half" indexes are directly associated to the $\frac{1}{2}$ factor in (C.9), thus producing (C.11). Of course, "half" indexes are implicit in computer implementation of Fig. C.5, i.e., space loop between lines 99 and 101 advances electric field from time t (from previous time iteration) to time $t + \Delta_t$, while space loop between lines 113 and 115 advances magnetic field from time $t - \frac{1}{2}\Delta_t$ (from previous time iteration) to time $t + \frac{1}{2}\Delta_t$. Observe that $H_y^{n+1/2}$ and E_z^{n+1} are obtained using fields previously computed.

```
/*
  Prof. Dr. Rodrigo Melo e Silva de Oliveira
  Universidade Federal do Pará (UFPA)

# FINITE-DIFFERENCE TIME-DOMAIN (FDTD) - DIDACTIC EXAMPLE

  FDTD 1D & MUR ABC

  05/10/2008
*/

// COMPILE WITH: gcc -lm -O

#include <stdio.h>
#include <stdlib.h>
#include <math.h>

// // // // // // // // // // //
#define Tsim (10E-6) /* Total simulation time (s) */
#define nx 8000   /* Total number of points in space */

// UNIVERSAL CONSTANTS:
#define mu0  1.25663706E-6
#define eps0 8.85418782E-12
#define C0   299792458E0
#define pi   3.14159265358979323846264338327 9E0
// // // // // // // // // //

int main(){
      int i,n,l;
      double dfdx;

      double ez[nx], hy[nx];   //FIELDS
      double eps[nx];

      double dx = 0.1;  // SPATIAL STEP (m)

      double t;  // TIME (s)
      double dtc = (dx/C0); //COURANT LIMIT (s)
      double dt = 0.99E0*dtc; //TIME STEP (s)
      printf("%e\n",dt);
      int nt = (int) round(Tsim/dt);  // TOTAL NUMBER OF TIME STEPS

 double T = 100.0*dt; // T: ASSOCIATED TO TIME DURATION OF EXCITATION
 SOURCE (s)

//  INITIALIZATION OF FIELDS
      for(i=0;i<nx;i++){
        ez[i] = 0.0E0;
        hy[i] = 0.0E0;
      }

// DIELECTRIC DEFINITION (FREE SPACE)
      for (i=0;i<nx;i++){
          eps[i] = eps0;
      }

//     EXAMPLE OF DIELECTRIC BLOCK (EPSr = 10)
   int ibi = 5000;
    int ibf = ibi + 960;

      for (i=ibi+1 ; i<ibf ; i++){
          eps[i] = 10.0E0*eps0;
       }
       eps[ibi] = 11.0E0*eps0/2.0E0;
       eps[ibf] = 11.0E0*eps0/2.0E0;
```

Fig. C.4 Didactic example of FDTD-1D implementation: definition of basic variables and parameters

```

    int ifonte = ibi-1000;  // SOURCE POSITION
//      int ifonte = nx/2;

    double F;

//  SENSOR FILE
    FILE *sensor;
     sensor = fopen("SENSOR.dat","w");
     rewind(sensor);

     int isensor = ibf + 10; // EXAMPLE OF SENSOR POSITION

// // // // // // // // // // // // // // // // // // // //
// MUR PARAMETERS
   int nxm = nx-1;
   double ez1n;
   double ez1nr;
//     double CMur = (C0*dt-dx)/(C0*dt+dx);
   double CMur = (1.0E0/dx - 1.0E0/(C0*dt))/(1.0E0/dx + 1.0E0/(C0*dt));
// // // // // // // // // //

// // FDTD  //  //
//***********/
  t = 0.0E0;  n=0;   // INITIALIZATION OF TIME

  while (n <= nt){ // TIME LOOP
    n++;  t += dt;   // TIME INCREMENTS

//********* ELECTRIC FIELD E   ************************
    ez1n = ez[1];
    ez1nr = ez[nxm-1];

    for(i=1;i<=nxm-1;i++){
      ez[i] = ez[i] + (dt/eps[i])*(   ( hy[i]-hy[i-1] )/dx  );
    }// E

//         MUR ABSORBING BOUNDARY CONDITIONS (1ST ORDER ABC)
      ez[0]   = ez1n + (CMur)*( ez[1] - ez[0] );          // LEFT SIDE
      ez[nxm] = ez1nr + (CMur)*( ez[nxm-1] - ez[nxm] );  // RIGHT SIDE

//*******  HARD SOURCE - E FILED EXCITATION ****************
  F =  exp( - pow (   ( ( t-3.0e0*T) / T ) ,  2.0E0)   );
//  if(t < (1.30E-15) )
   ez[ifonte] = F; // HARD SOURCE

//********* MAGNETIC FIELD H   ************************
 for(i=0;i<=nxm-1;i++){ // campo H
         hy[i] = hy[i] + (dt/mu0)*(   ( ez[i+1]-ez[i] )/dx  );
    } // H

//   ********  sensor (E-field record)   **************
  fprintf( sensor,"%e %e\n", t, ez[isensor] );

} //END OF TIME LOOP

  fclose(sensor);

} //main
```

Fig. C.5 Didactic example of FDTD-1D implementation: additional variables and the main FDTD algorithm implementation

In this work, source is implemented by forcing electric field to be governed, at the excitation point, by a explicitly defined function $F(t)$, as it is illustrated in Fig. C.3. Computer implementation can be seen in Fig. C.5 (lines 108–110), in which an illustrative Gaussian pulse is calculated at time $t = n\Delta_t$ and directly attributed to E_z at the excitation point. This attribution must take place after Ez is updated using Ampere's law in all analysis space (see lines 99–101). This method of excitation, which corresponds to apply a hard time-variant voltage source by controlling electric field in space, was used validate our laboratory experiments of Figs. 5.15, 5.16 and 5.21 by conceiving the FDTD-3D model of Fig. 5.22. Function $F(t)$ was defined according to the output voltage of the artificial partial discharge instrumentation output (see Fig. 5.21a, the injected signal). Voltage source $V_s(t)$ and E_z at the source point (x_s, y_s, z_s), in FDTD three-dimensional space, are related by $E_z(t, x_s, y_s, z_s) = V_s(t)/\Delta_z$.

As it can be observed in the described FDTD technique, space must be truncated. Frequently, it is required that the electromagnetic waves are absorbed at the ends of computational domain for simulating open space with finite number of cells. A simple absorbing boundary condition (ABC) is that described by Mür in [23]. Assuming that, in a one-dimensional space, electromagnetic wave is excited and propagates at a region in which $x > 0$, Mür demonstrates that outward waves can be approximately be described, for our one-dimensional problem defined in Fig. C.3, by the first order equation

$$\frac{\partial E_z}{\partial x} - \frac{1}{C}\frac{\partial E_z}{\partial t} = 0, \tag{C.16}$$

in which C is wave speed. We can write a discrete version of (C.16) by using central differences to approximate derivatives. However, since (C.16) describes E_z in terms of spatial and temporal derivatives of itself (magnetic field is not explicitly considered), time synchronization and spatial centering must be guaranteed. This can be achieved by time-averaging discrete versions of $\partial E_z/\partial x$ and space-averaging discrete versions of $\partial E_z/\partial t$, as it has been done in

$$\begin{aligned}&\left(\frac{E_z^n(i) - E_z^n(i-1)}{\Delta x} + \frac{E_z^{n+1}(i) - E_z^{n+1}(i-1)}{\Delta x}\right)\frac{1}{2}\\ &\quad - \frac{1}{C}\left(\frac{E_z^{n+1}(i) - E_z^n(i)}{\Delta t} + \frac{E_z^{n+1}(i-1) - E_z^n(i-1)}{\Delta t}\right)\frac{1}{2} = 0,\end{aligned} \tag{C.17}$$

which produces, after proper mathematical manipulations,

$$\begin{aligned}&\left(\frac{1}{\Delta x} + \frac{1}{C \cdot \Delta t}\right)E_z^n(i) - \left(\frac{1}{\Delta x} - \frac{1}{C \cdot \Delta t}\right)E_z^n(i-1)\\ &\quad + \left(\frac{1}{\Delta x} - \frac{1}{C \cdot \Delta t}\right)E_z^{n+1}(i) - \left(\frac{1}{\Delta x} + \frac{1}{C \cdot \Delta t}\right)E_z^{n+1}(i-1) = 0.\end{aligned} \tag{C.18}$$

For the left-side boundary, by setting $i = 1$, and rearranging terms, we see from (C.18) that

$$E_z^{n+1}(0) = E_z^n(1) - \frac{\left(\frac{1}{\Delta x} - \frac{1}{C \cdot \Delta t}\right)}{\left(\frac{1}{\Delta x} + \frac{1}{C \cdot \Delta t}\right)} E_z^n(0) + \frac{\left(\frac{1}{\Delta x} - \frac{1}{C \cdot \Delta t}\right)}{\left(\frac{1}{\Delta x} + \frac{1}{C \cdot \Delta t}\right)} E_z^{n+1}(1). \tag{C.19}$$

Implementation of (C.19) is seen in line 104 in the computer program of Fig. C.5. Mür equation seen in line 105, for absorbing outward waves at the right side of the computational domain, is obtained analogously to the procedure developed to obtain (C.19).

Finally, for sake of completeness, we present FDTD-3D equations obtained by employing central differences to time and space derivatives in (C.3)–(C.8). The equations used in this work to obtain fields in the three-dimensional FDTD domains are

$$H^{n+\frac{1}{2}}_{x\ (i,j+\frac{1}{2},k+\frac{1}{2})} = H^{n-\frac{1}{2}}_{x\ (i,j+\frac{1}{2},k+\frac{1}{2})} +$$

$$+ \frac{\Delta_t}{\mu}\left[\frac{E^n_{y\ (i,j+\frac{1}{2},k+1)} - E^n_{y\ (i,j+\frac{1}{2},k)}}{\Delta_z} - \frac{E^n_{z\ (i,j+1,k+\frac{1}{2})} - E^n_{z\ (i,j,k+\frac{1}{2})}}{\Delta_y}\right], \tag{C.20}$$

$$H^{n+\frac{1}{2}}_{y\ (i+\frac{1}{2},j,k+\frac{1}{2})} = H^{n-\frac{1}{2}}_{y\ (i+\frac{1}{2},j,k+\frac{1}{2})} +$$

$$+ \frac{\Delta_t}{\mu}\left[\frac{E^n_{z\ (i+1,j,k+\frac{1}{2})} - E^n_{z\ (i,j,k+\frac{1}{2})}}{\Delta_x} - \frac{E^n_{x\ (i+\frac{1}{2},j,k+1)} - E^n_{x\ (i+\frac{1}{2},j,k)}}{\Delta_z}\right], \tag{C.21}$$

$$H^{n+\frac{1}{2}}_{z\ (i+\frac{1}{2},j+\frac{1}{2},k)} = H^{n-\frac{1}{2}}_{z\ (i+\frac{1}{2},j+\frac{1}{2},k)} +$$

$$+ \frac{\Delta_t}{\mu}\left[\frac{E^n_{x\ (i+\frac{1}{2},j+1,k)} - E^n_{x\ (i+\frac{1}{2},j,k)}}{\Delta_y} - \frac{E^n_{y\ (i+1,j+\frac{1}{2},k)} - E^n_{y\ (i,j+\frac{1}{2},k)}}{\Delta_x}\right], \tag{C.22}$$

$$E^{n+1}_{x\ (i+\frac{1}{2},j,k)} = E^n_{x\ (i+\frac{1}{2},j,k)}\left(\frac{1 - \sigma\frac{\Delta_t}{2\epsilon}}{1 + \sigma\frac{\Delta_t}{2\epsilon}}\right) +$$

$$+\frac{\Delta_t}{\epsilon\left(1+\sigma\frac{\Delta_t}{2\epsilon}\right)}\left[\frac{H^{n+\frac{1}{2}}_{z\ (i+\frac{1}{2},j+\frac{1}{2},k)}-H^{n+\frac{1}{2}}_{z\ (i+\frac{1}{2},j-\frac{1}{2},k)}}{\Delta_y}\right]$$

$$-\frac{\Delta_t}{\epsilon\left(1+\sigma\frac{\Delta_t}{2\epsilon}\right)}\left[\frac{H^{n+\frac{1}{2}}_{y\ (i+\frac{1}{2},j,k+\frac{1}{2})}-H^{n+\frac{1}{2}}_{y\ (i+\frac{1}{2},j,k-\frac{1}{2})}}{\Delta_z}\right], \quad \text{(C.23)}$$

$$E^{n+1}_{y\ (i,j+\frac{1}{2},k)}=E^{n}_{y\ (i,j+\frac{1}{2},k)}\left(\frac{1-\sigma\frac{\Delta_t}{2\epsilon}}{1+\sigma\frac{\Delta_t}{2\epsilon}}\right)+$$

$$+\frac{\Delta_t}{\epsilon\left(1+\sigma\frac{\Delta_t}{2\epsilon}\right)}\left[\frac{H^{n+\frac{1}{2}}_{x\ (i,j+\frac{1}{2},k+\frac{1}{2})}-H^{n+\frac{1}{2}}_{x\ (i,j+\frac{1}{2},k-\frac{1}{2})}}{\Delta_z}\right]$$

$$-\frac{\Delta_t}{\epsilon\left(1+\sigma\frac{\Delta_t}{2\epsilon}\right)}\left[\frac{H^{n+\frac{1}{2}}_{z\ (i+\frac{1}{2},j+\frac{1}{2},k)}-H^{n+\frac{1}{2}}_{z\ (i-\frac{1}{2},j+\frac{1}{2},k)}}{\Delta_x}\right], \quad \text{(C.24)}$$

and

$$E^{n+1}_{z\ (i,j,k+\frac{1}{2})}=E^{n}_{z\ (i,j,k+\frac{1}{2})}\left(\frac{1-\sigma\frac{\Delta_t}{2\epsilon}}{1+\sigma\frac{\Delta_t}{2\epsilon}}\right)+$$

$$+\frac{\Delta_t}{\epsilon\left(1+\sigma\frac{\Delta_t}{2\epsilon}\right)}\left[\frac{H^{n+\frac{1}{2}}_{y\ (i+\frac{1}{2},j,k+\frac{1}{2})}-H^{n+\frac{1}{2}}_{y\ (i-\frac{1}{2},j,k+\frac{1}{2})}}{\Delta_x}\right]$$

$$-\frac{\Delta_t}{\epsilon\left(1+\sigma\frac{\Delta_t}{2\epsilon}\right)}\left[\frac{H^{n+\frac{1}{2}}_{x\ (i,j+\frac{1}{2},k+\frac{1}{2})}-H^{n+\frac{1}{2}}_{x\ (i,j-\frac{1}{2},k+\frac{1}{2})}}{\Delta_y}\right]. \quad \text{(C.25)}$$

Grid truncation can be performed with Mür formulation [23] or with PMLs (Perfect Matched Layers) [26, 43, 45], which produce higher levels of absorption than Mür ABCs [43].

Appendix D
Linear Time-Invariant Filtering

This appendix gives a short introduction to the topic of linear time-invariant filtering. We start by recalling the concepts of linearity and invariance of systems. Then, we present the idea of filtering associated with the properties of selectivity in the frequency domain. Design examples of most common approaches conclude the appendix.

D.1 Linearity and Time-Invariance

Linear systems are those to which the superposition theorem applies, i.e., Eq. (D.1) holds.[1]

$$T\left\{\alpha x_1[n] + \beta x_2[n]\right\} = \alpha T\left\{x_1[n]\right\} + \beta T\left\{x_2[n]\right\}, \tag{D.1}$$

where $T\{\cdot\}$ represents the system response (transformation imposed) to an input signal; α and β are real constants; and x_1 and x_2 are input signals.

In other words, the system output of a certain linear combination of input signals is the linear combination (same weighting constants) of the corresponding outputs (each output due to only one of the input signals, separately).

The other property of interest, is the time-invariance, defined according to the following relations:

$$y[n] = T\left\{x[n]\right\}, \tag{D.2}$$

$$T\left\{x[n-k]\right\} = y[n-k] \qquad \text{with } k \in \mathbb{Z}. \tag{D.3}$$

[1] We consider the signal processing chain is implemented in the discrete domain and our notation for theorems, system properties and signal transforms will reflect this.

V. Dmitriev et al., *Partial Discharges in Hydroelectric Generators*, Power Systems,
https://doi.org/10.1007/978-3-031-36604-8

Equations (D.2) and (D.3) state that, for a given pair of input/output signals ($x[n]$ and $y[n]$, respectively), a shift of k samples in the input ($x[n-k]$) of a time-invariant system will cause an equal shift in the output ($y[n-k]$) and this will be the only output characteristic changed by the input shift.

As a consequence, the combination of linearity and time-invariance properties into one single system allow us to represent any output of that system by the convolution between the input signal and its impulse response:

$$y[n] = \sum_{k=-\infty}^{\infty} h[k]x[n-k], \tag{D.4}$$

$$= h[n] * x[n], \tag{D.5}$$

where $*$ indicates the convolution operation and $h[n]$ denotes the impulse response of the LTI system.

We can also represent the impulse response as

$$h[n] = T\{\delta[n]\}, \tag{D.6}$$

where $\delta[n]$ is the unit sample sequence, also commonly called unit impulse signal, whose definition is

$$\delta[n] = \begin{cases} 1, & n = 0 \\ 0, & n \neq 0. \end{cases} \tag{D.7}$$

From Eqs. (D.4) and (D.5), we can see that if the $h[n]$ of a LTI system is known, one can determine exactly the response of such a system to any input signal $x[n]$. In this sense, the impulse response fully represents a LTI system. Additional and important insights about how a LTI system affects input signals are provided by its frequency response, which is defined as the Fourier transform of the impulse response:

$$H(\omega) = \mathcal{F}\{h[n]\}, \tag{D.8}$$

$$= \sum_{n=-\infty}^{\infty} h[n]e^{-j\omega n}, \tag{D.9}$$

where $\mathcal{F}$ is the Fourier transform operator; and ω denotes frequency (actually, a normalized frequency, given in rad).

The frequency response $H(\omega)$ is a complex function of ω and explains the way the system impacts on the spectral components of the input signal. Specifically, the output magnitude $|Y(\omega)|$ is a weighted version of the magnitude of $X(\omega)$, in which $|H(\omega)|$ acts as the weighting function. Likewise, the output phase $\angle Y(\omega)$ is given by the input phase $\angle X(\omega)$ added to the frequency response phase of the system $\angle H(\omega)$.

This can be better understood by recalling the convolution property of the Fourier transform:

$$\mathcal{F}\{y[n]\} = \mathcal{F}\{h[n] * x[n]\}, \tag{D.10}$$

$$Y(\omega) = H(\omega)X(\omega), \tag{D.11}$$

$$|Y(\omega)|e^{j\angle Y(\omega)} = |H(\omega)||X(\omega)|e^{j\left[\angle H(\omega)+\angle X(\omega)\right]}. \tag{D.12}$$

D.2 Selectivity in the Frequency Domain

The success of using LTI filtering for denoising and reduction of interference is conditioned by the existence of a certain degree of spectral separation between signal and unwanted components.

The general idea is designing a LTI filter whose frequency response removes spectral regions corresponding to the undesired signals (interference or noise). Figure D.1 shows a general template for the frequency response magnitude of a bandpass filter.[2] Definition of filter parameters (δ_p, δ_s, f_{s1}, f_{s1}, f_{p1}, and f_{p2}) requires a specialized knowledge about the application domain.

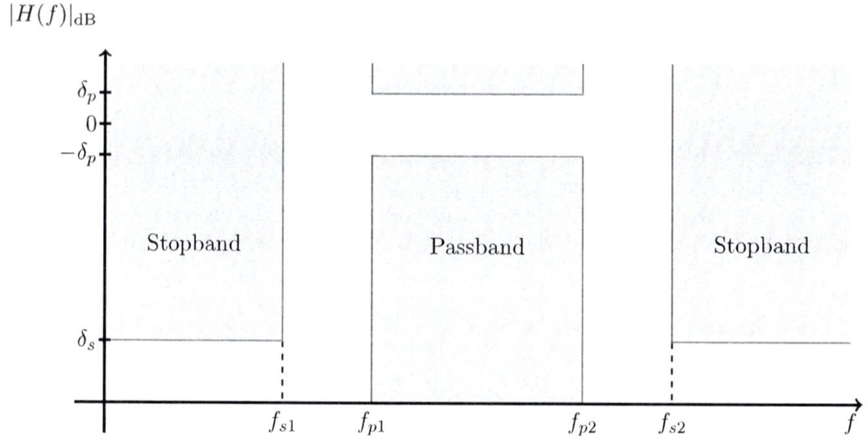

Fig. D.1 Specifications for the frequency response magnitude $|H(f)|_{\mathrm{dB}}$ of a bandpass filter: δ_p and δ_s denote gain variation at passband and the maximum gain at stopband regions, respectively; while f_s and f_p are stopband and passband frequencies, respectively

[2]Other types of filter, namely highpass, lowpass, reject-band, and notch, are not discussed in this text. A thorough introduction to linear filtering would lengthen our text without significant contribution to the main subject (partial discharge denoising). We encourage the reader interested in going deeper into linear filtering to refer to [46].

The passband corresponds to the frequency interval whose input components are intended to be preserved, being associated to where the signal of interest (in our case, PD pulses) is located in the spectral domain. The passband is limited by the frequencies f_{p1} and f_{p2}. In turn, the stopband indicates frequency regions to be suppressed from the input signal, because dominated by either noise or interference. The stopband is defined by the frequencies f_{s1} and f_{s2}. The transition band (gaps between passband and stopband) specifies how selective a filter is: the more selective is a filter (i.e., the narrower its transition band), the higher is its order. For a digital filter, the order is directly related to computational complexity and memory concerns. The parameter δ_p specifies the allowed gain variation in the passband, while δ_s determines the maximum gain in stopband zones.

Any filter whose frequency response magnitude $|H(f)|$ does not hit the gray area in Fig. D.1 is valid as it complies with a given set of specifications.

D.3 Finite Impulse Response (FIR) Filters

The most important characteristic of a FIR filter is that its stability is guaranteed, because a finite-length $h[n]$ always respects the stability condition for LTI systems:

$$\sum_{n=-\infty}^{\infty} |h[n]| < \infty. \tag{D.13}$$

This appendix discusses two commonly used approaches to design FIR filters: windowing [47] and Park-McClellan [48] methods.

The starting point of the windowing method is the impulse response of an ideal filter that follows the intended design specifications. Then, to obtain a causal FIR filter, we shift the ideal impulse response and truncate it by using a finite-length window:

$$h_w[n] = h_i[n]w[n], \tag{D.14}$$

where $h_w[n]$, $h_i[n]$, and $w[n]$ denote the designed filter impulse response, the shifted ideal (or desired) filter impulse response, and the weighting window, respectively.

The shape and length of $w[n]$ determine the characteristics of the resulting frequency response. The multiplication property of Fourier transform provides the theoretical framework to understand changes in ideal filter features due to windowing:

$$H_w(\omega) = \mathcal{F}\{h_i[n]w[n]\}, \tag{D.15}$$

$$= \frac{1}{2\pi} \int_{-\pi}^{\pi} H_i(\theta) W(\omega - \theta)\, d\theta, \tag{D.16}$$

where $H_w(\omega)$, $H_i(\omega)$ and $W(\omega)$ are the Fourier transforms of $h_w[n]$, $h_i[n]$ and $w[n]$, respectively.

Note that Eq. (D.16) shows the designed frequency response $H_w(\omega)$ is given by periodic convolution between the ideal/desired frequency response $H_i(\omega)$ and the Fourier transform of the window function $W(\omega)$. Next, we present two standard weighting functions to demonstrate how windows affect the frequency response of the resulting FIR filter. We begin by the well-known rectangular window, whose definition in the sample domain is as follows:

$$w[n] = \begin{cases} 1, & 0 \leq n \leq N \\ 0, & \text{otherwise,} \end{cases} \tag{D.17}$$

where N is an integer.

As a second example, we take the popular Hann window, expressed in (D.18).

$$w[n] = \begin{cases} 0.5 - 0.5\cos\left(\frac{2\pi n}{N}\right), & 0 \leq n \leq N \\ 0, & \text{otherwise.} \end{cases} \tag{D.18}$$

Figure D.2a, b compare both windows for $N = 9$ in the sample and frequency domains, respectively. An important aspect to notice is how the shape of the window in the sample domain modifies the side-lobes in the frequency domain. Because the Hann window is smoother than the rectangular window (Fig. D.2a), the side-lobes of the former have a smaller magnitude than the rectangular's (Fig. D.2b). This characteristic is important, as we will see soon, to control the maximum gain in the stopband.

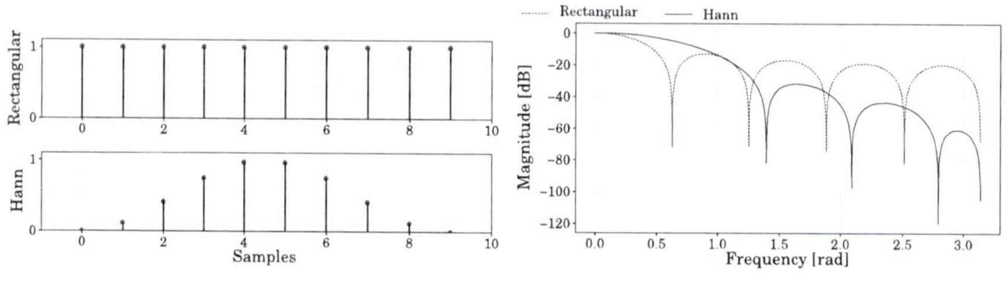

(a) (b)

Fig. D.2 Comparison between the rectangular and Hann window functions in (**a**) sample and (**b**) frequency domains. Both windows are calculated with $N = 9$

The window method is direct and simple, but offers very little control over the resulting filter frequency response. Two aspects of the filter frequency response are influenced by the window shape: transition band width (i.e., the distance between stopband and passband frequencies), and the minimum attenuation at the stopband.

The transition band is inversely related to main lobe width of the window Fourier transform: the larger the main lobe, the larger the transition band of the resulting filter. By duality, one can narrow the main lobe width by increasing the window length and vice-versa. For the same N, note the rectangular window has a narrower main lobe when compared with the one of the Hann window (Fig. D.2b). As a result, the order of the FIR filter obtained with the rectangular window is lower than the one with the Hann window for a given transition band.

In turn, the minimum attenuation at the stopband region is determined by the ratio between the maximum of the main and of the first side lobes: the smaller the first side lobe amplitude with respect to the main lobe amplitude, the higher the minimum stopband attenuation (or equivalently, the smaller the maximum gain in the stopband). Such a ratio does not depend on the length of the window, but on its shape (Fig. D.2).

We have designed two FIR filters by using the rectangular and Hann windows, considering the specifications given in Table D.1.

Figure D.3 shows the frequency response of those filters. Compare the length (L) and the stopband gains for each window. Note the order of the filter designed with Hann window is approximately twice the order of the rectangular window. On the other hand, stopband attenuation is almost 30 dB higher for the Hann-window filter.

Table D.1 Hypothetical bandpass filter specifications. The chosen values are consistent with the case of PD monitoring in HF/VHF range

Design parameter	Value
δ_p	1 dB
δ_s	−20 dB
f_{s1}	50 MHz
f_{p1}	60 MHz
f_{p2}	300 MHz
f_{s2}	310 MHz

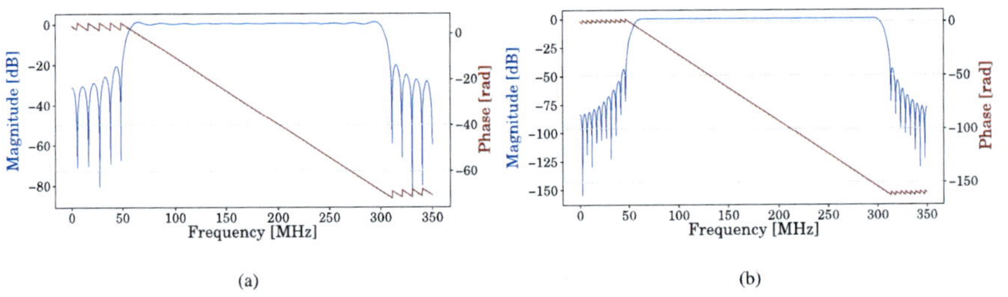

Fig. D.3 Frequency response (magnitude and phase) of FIR filters designed by the window method: (**a**) rectangular window ($L = 66$); (**b**) Hann window ($L = 139$)

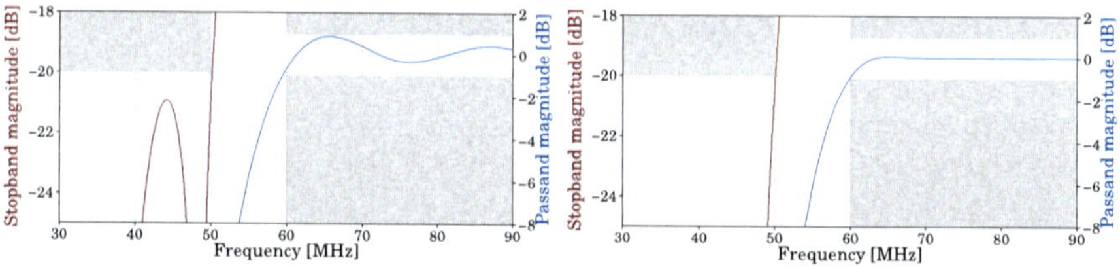

(a)

(b)

Fig. D.4 Detailed view of the left transition band for the FIR filters designed by the window method: (**a**) rectangular window ($L = 66$); (**b**) Hann window ($L = 139$). To observe design specifications, the magnitude of frequency response must not hit the gray areas

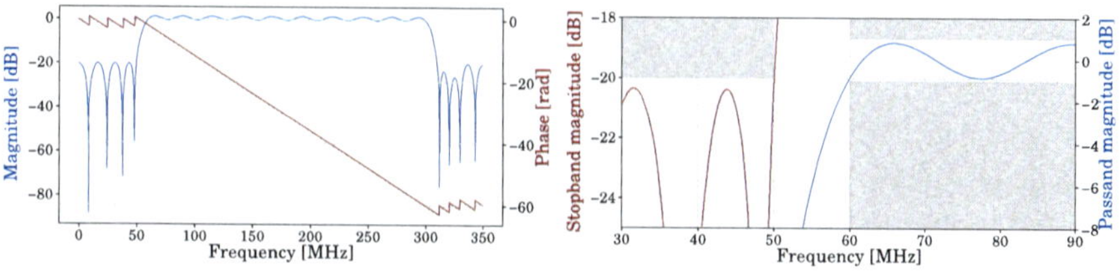

(a)

(b)

Fig. D.5 Frequency response of a FIR filter designed by the Park-McClellan algorithm ($L = 55$): (**a**) magnitude and phase; (**b**) left transition band details

Figure D.4 gives us additional details about the left transition band of each designed filter. Although different, both filters comply with design specifications. Also note the "oscillatory" behavior in the passband of the rectangular-window filter and again the higher attenuation in the stopband of the Hann-window filter.

The next method for designing FIR filters we address is the Parks-McClellan procedure [48]. This method calculates the filter coefficients by optimizing an approximation error function between the desired and the actual frequency responses:

$$E(\omega) = W(\omega)\,[H_d(\omega) - A_e(\omega)]\,, \tag{D.19}$$

where $E(\omega)$ is the approximation error function; $W(\omega)$ is a general weighting function; $H_d(\omega)$ represents the desired frequency response; and $A_e(\omega)$ denotes the approximating function, which depends on the filter impulse response.

Figure D.5 depicts the frequency response of a FIR filter designed by the Park-McClellan procedure with impulse response length $L = 55$. In this case, we can see the "oscillatory" behavior of the frequency response in the passband and attenuation levels that barely respect stopband restrictions. When comparing with the two previous FIR designs, it seems the Park-McClellan filter has taken the worst of them. However, there is a sort of compensation regarding resulting order $L = 55$, which is the smallest among our examples so far.

D.4 Infinite Impulse Response (IIR) Filters

Let us move to another class of LTI filters: those with infinite impulse response (IIR filters). Such a feature is attained by signal feedback paths in the implementation structure. An advantage of an IIR filter is that its order is generally much smaller than an equivalent FIR filter (i.e., FIR and IIR filters designed to comply the same specifications). The downside is that feedback loops are always followed by instability concerns.

The most common strategies used for discrete IIR filter synthesis are based on the well-known design techniques developed for continuous IIR systems. A typical IIR design workflow comprises: (a) analog IIR synthesis by using a chosen approximation function; and (b) mapping the resulting analog transfer function to a discrete transfer function.

In the following, we present four examples of IIR systems tuned to the bandpass filter specified in Table D.1. Each example refers to a different approximation function: Buttherworth's, Chebyshev's (type I and II) and Cauer's (Elliptic function). We emphasize their main trade-offs in terms of order and frequency response behavior. In all cases, we adopt bilinear transformation [47] to map from continuous to discrete domains.

The magnitude of the continuous low-pass Butterworth filter gain is given by

$$|H(\Omega)| = \left[1 + \varepsilon^2 \left(\frac{\Omega}{\Omega_p}\right)^{2n}\right]^{-\frac{1}{2}}, \tag{D.20}$$

where Ω represents angular frequency, ε is a constant, Ω_p denotes the passband frequency, and n is the order of the Butterworth function.

In Eq. (D.20), Ω_p, ε and n are the parameters to be defined based on the set of filter specifications. A passband version of Eq. (D.20) can be obtained by using frequency transformation techniques [46].

In Fig. D.6, one can see the designed Butterworth frequency response. This approximation function is characterized by a frequency magnitude that is monotonic, maximally flat (its slope is as flat as possible at $\Omega = 0$) and smooth. Note the order of this filter ($n = 31$) is significantly lower than FIR filters in the previous section. However, as we will see soon, the smoothness of the Butterworth filter is associated to a higher order when comparing with other IIR approximation functions. Also note the non-linear phase of the passband. Figure D.6b reveals details regarding the left transition band, where monotonicity of a Butterworth frequency response is apparent.

The next IIR filter we are about to address is the Chebyshev type I filter, whose gain function is as follows

$$|H(\Omega)| = \left[1 + \varepsilon^2 C_n^2 \left(\frac{\Omega}{\Omega_p}\right)\right]^{-\frac{1}{2}}, \tag{D.21}$$

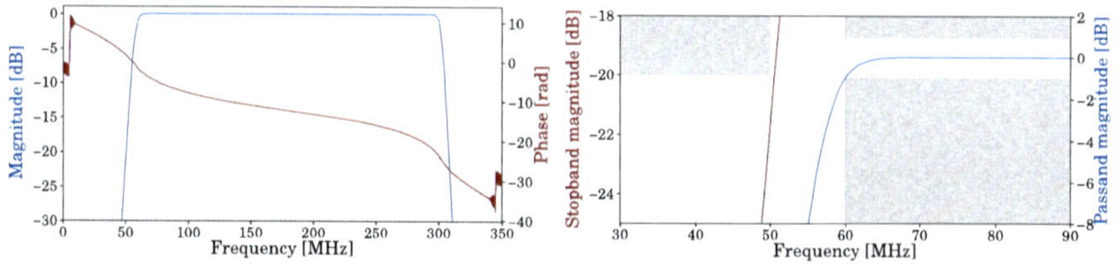

(a) (b)

Fig. D.6 Frequency response of a IIR filter designed with the Butterworth approximation function ($n = 31$): (**a**) magnitude and phase of the frequency response; (**b**) left transition band details

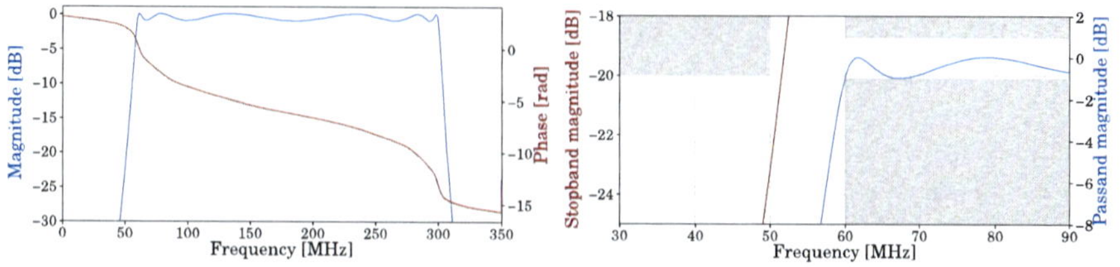

(a) (b)

Fig. D.7 Frequency response of a IIR filter designed with the Chebyshev (type I) approximation function ($n = 13$): (**a**) magnitude and phase of the frequency response; (**b**) left transition band details

where $C_n(\Omega)$ is the n-order Chebyshev function, which is defined as

$$C_n(\Omega) = \begin{cases} \cos\left(n \cos^{-1}\Omega\right) & |\Omega| \leq 1 \\ \cosh\left(n \cosh^{-1}\Omega\right) & |\Omega| > 1. \end{cases} \tag{D.22}$$

Figure D.7 shows the Chebyshev type I filter designed according to our hypothetical specifications. Comparing with the equivalent Butterworth filter, the Chebyshev type I filter has lower order ($n = 13$) at the expense of the monotonicity of the frequency response in the passband. Figure D.7b details the left transition band, in which the typical oscillatory behavior of Chebyshev type I filters in the passband is evident.

The third approximation function is the Chebyshev type II, which is based on a modification of the gain function in Eq. (D.21):

$$|H(\Omega)| = \left[1 + \frac{1}{\varepsilon^2 C_n^2\left(\frac{\Omega}{\Omega_p}\right)}\right]^{-\frac{1}{2}}. \tag{D.23}$$

Regarding the Chebyshev type I, Eq. (D.23) exhibits a switch between flat and oscillatory sections of the frequency response. This can be seen in Fig. D.8, where

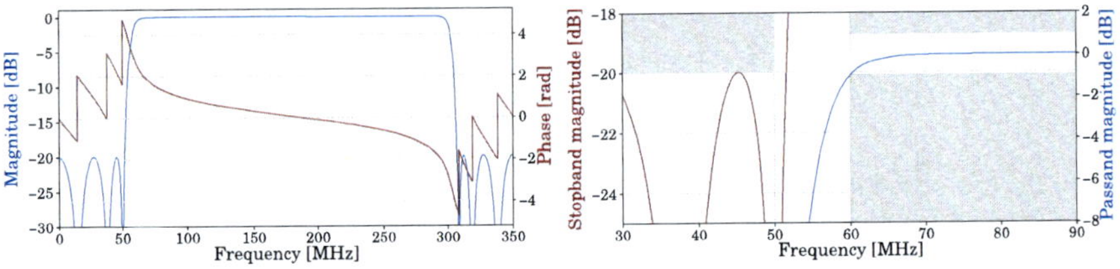

(a) (b)

Fig. D.8 Frequency response of a IIR filter designed with the Chebyshev (type II) approximation function ($n = 13$): (**a**) magnitude and phase of the frequency response; (**b**) left transition band details

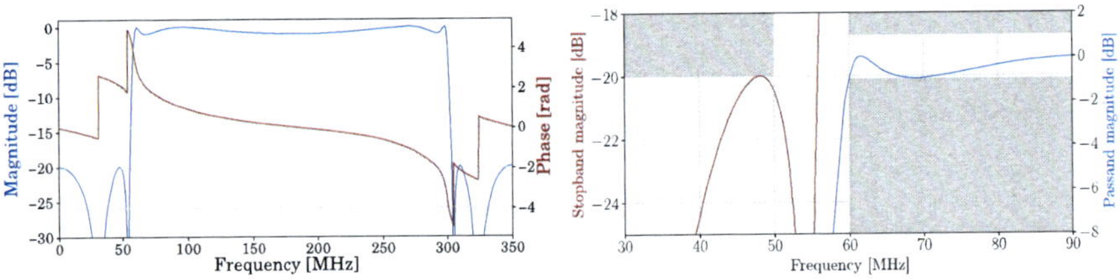

(a) (b)

Fig. D.9 Frequency response of a IIR filter designed with the Elliptic approximation function ($n = 9$): (**a**) magnitude and phase of the frequency response; (**b**) left transition band details

the frequency response of the designed Chebyshev filter type II is shown. The order of the filter is the same of the Chebyshev type I ($n = 13$).

Our fourth IIR filter example is the Elliptic (a.k.a., Cauer) filter. One distinct characteristic of this filter is the presence of oscillatory behavior both in passband and stopband (Fig. D.9). Such a feature is partially made up for a filter with lower order than its Chebyshev and Butterworth counterparts. The order of the Elliptic filter, whose frequency response is given in Fig. D.9, is $n = 9$. The mathematical theory related to the Elliptic filters is out of the scope of this text. The interested reader should refer to [46] for a detailed treatment of the subject.

D.5 Summary of LTI Filter Characteristics

Table D.2 summarizes the essential differences between FIR and IIR filter approaches, regarding stability, and system order. About stability, it suffices to say that due to its finite length, the impulse response of a FIR filter is always absolutely summable, thus stable, while the stability of IIR filters must be double checked: first time after design and then after implementation. The second stability verification is justified because numerical issues in parameter representation may eventually move any or some of the system poles to the unstable region of the

Table D.2 Comparison of general characteristics between FIR and IIR filters

Feature	FIR	IIR
Stability	Always stable	Not guaranteed
Order	Higher	Lower
Computational complexity	Higher	Lower
Required memory speed	Higher	Lower
Required memory size	Higher	Lower
Required processing power	Higher	Lower

Table D.3 Filter orders and frequency response behavior (SB: stopband; PB: passband)

Class	Approximation/Technique	Order	Oscillatory behavior
FIR	Rectangular windowing	66	SB/PB
FIR	Hann windowing	139	SB/PB
FIR	Parker-McClellan algorithm	55	SB/PB
IIR	Butterworth	31	–/–
IIR	Chebyshev I	13	–/PB
IIR	Chebyshev II	13	SB/–
IIR	Elliptic	9	SB/PB

complex plane. System order, in turn, is related to the computational complexity required for filter output calculations as well as to the cost of the implementation (required memory units, computational power, etc.).

Table D.3 lists orders and frequency response behavior of the designed filters presented so far. Note the order of IIR filters are, in our examples, at least half of the order of FIR filters. It is also worthwhile to observe the variation of order among IIR filters.

Appendix E
An Introduction to Wavelets

This appendix focuses on techniques for denoising applied to signals represented in the wavelet domain. We present a brief review of the main concepts on the subject, followed by the most common types of wavelet, implementation schemes by using filter banks, and primary thresholding strategies to reduce noise. One reason why wavelets are so important today is they provide more flexibility than Fourier and short Fourier transforms regarding the definition of kernels used to represent time-frequency information of signals [49].

E.1 Multiresolution Approximation

Multiresolution approximation of functions is a concept closely related to wavelets. We begin this topic by stating we can construct a family of orthonormal functions $\psi_{j,k}(t)$ by translating and dilating a special function $\psi(t)$, called *mother wavelet*:

$$\left\{ \psi_{j,k}(t) = \frac{1}{\sqrt{2^j}} \psi\left(\frac{t - 2^j k}{2^j}\right) \right\}_{(j,k) \in \mathbb{Z}^2}, \tag{E.1}$$

where j and k are integer parameters that control the dilation and translation of the function $\psi(t)$, respectively.

The functions $\psi_{j,k}$ in (E.1) form a basis for $\mathbf{L}^2(\mathbb{R})$, i.e., the space of functions $f : \mathbb{R} \to \mathbb{R}$, for which

$$\int_{-\infty}^{\infty} |f(t)|^2 dt < \infty. \tag{E.2}$$

Hence, a function $f \in \mathbf{L}^2(\mathbb{R})$ can be expressed as a linear combination

V. Dmitriev et al., *Partial Discharges in Hydroelectric Generators*, Power Systems,
https://doi.org/10.1007/978-3-031-36604-8

$$f(t) = \sum_j \sum_k < f, \psi_{j,k} > \psi_{j,k}(t), \tag{E.3}$$

where $< f, \psi_{j,k} >$ represents the inner-product between f and $\psi_{j,k}$, in this context also called *wavelet coefficient*:

$$< f, \psi_{j,k} >= \int_{-\infty}^{\infty} f(t)\psi_{j,k}(t)dt. \tag{E.4}$$

The next point in this discussion is the notion of *resolution*, which is firmly linked to the representation of details: higher resolution corresponds to the capacity to reproduce finer details, while lower resolution implies coarser representation. Mathematically, the resolution is associated with subspaces at different scales j, $\mathbf{V}_j \subset \mathbf{L}^2(\mathbb{R})$. These subspaces are embedded, in the sense that a subspace at a higher scale (lower resolution) is always a subset of a subspace at a lower scale (higher resolution), i.e., $\mathbf{V}_i \subset \mathbf{V}_j$ for $i > j$ (Fig. E.1).

Assume we have a basis for $\mathbf{V}_j$, given by the following set of functions

$$\left\{\phi_{j,k}(t) = \frac{1}{\sqrt{2^j}}\phi\left(\frac{t-2^jk}{2^j}\right)\right\}_{(j,k)\in\mathbb{Z}^2}, \tag{E.5}$$

where the function ϕ (scaling function) is calculated from the mother wavelet ψ.

Now, we can approximate a function f for a given resolution:

$$f_j(t) = \sum_k < f, \phi_{j,k} > \phi_{j,k}(t), \tag{E.6}$$

where f_j is the approximation of f at the scale j.

The subspace $\mathbf{V}_j$ can be obtained by the subspace at the next scale $\mathbf{V}_{j+1}$

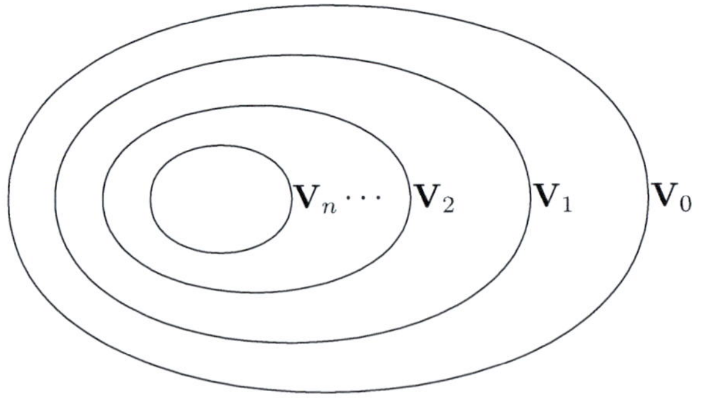

Fig. E.1 $\mathbf{V}_j$ represents a subspace at scale j. Finer details can be expressed at lower scales

$$\mathbf{V}_j = \mathbf{V}_{j+1} \oplus \mathbf{W}_{j+1}, \tag{E.7}$$

where $\mathbf{W}_{j+1}$ is the orthogonal complement of $\mathbf{V}_{j+1}$.

The fact that $\mathbf{W}_{j+1}$ is the orthogonal complement of $\mathbf{V}_{j+1}$ implies that:

(a) all functions $f_{j+1} \in \mathbf{V}_{j+1}$ are orthogonal to every function $w_{j+1} \in \mathbf{W}_{j+1}$, i.e., $< f_{j+1}, w_{j+1} >= 0$;

(b) a function in $\mathbf{V}_j$ can be expressed by a sum of functions in $\mathbf{V}_{j+1}$ and $\mathbf{W}_{j+1}$:

$$f_j(t) = f_{j+1}(t) + w_{j+1}(t), \tag{E.8}$$

where f_{j+1} and w_{j+1} are the approximation of f in $\mathbf{V}_{j+1}$ and $\mathbf{W}_{j+1}$, respectively:

$$f_{j+1}(t) = \sum_k < f, \phi_{j+1,k} > \phi_{j+1,k}(t), \tag{E.9}$$

$$w_{j+1}(t) = \sum_k < f, \psi_{j+1,k} > \psi_{j+1,k}(t). \tag{E.10}$$

In Eq. (E.8), the approximation of f at a higher resolution (f_j) is composed of a coarser approximation (f_{j+1}) to which details are added by the function w_{j+1}. We will return to this point when addressing filter banks to implement a wavelet decomposition.

E.2 Wavelet Types

Depending on the application, one should choose the type of wavelet that best expresses the shape and features of the signals of interest. Of the various types of wavelets, some of the most commonly used include [50, 51] (Fig. E.2):

- Haar wavelets:

$$\psi(t) = \begin{cases} -1, & \text{if } 0 \leq t < 1/2, \\ 1, & \text{if } 1/2 \leq t < 1, \\ 0, & \text{otherwise.} \end{cases} \tag{E.11}$$

- Morlet wavelets:

$$\psi(t) = e^{i\omega_0 t} e^{-t^2/2}, \tag{E.12}$$

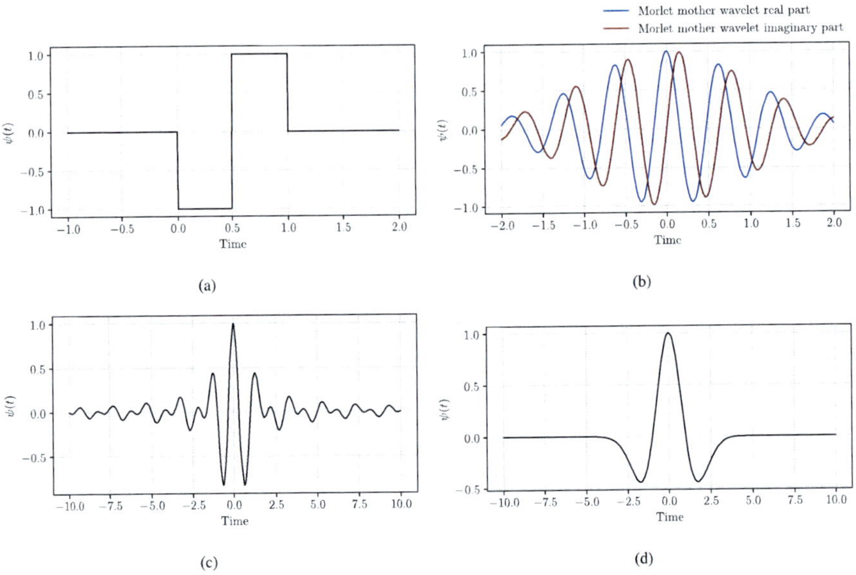

Fig. E.2 Some examples of different mother wavelets: (**a**) Haar mother wavelet; (**b**) Morlet mother wavelet; (**c**) Shannon mother wavelet; (**d**) Ricker mother wavelet

- Shannon wavelets:

$$\psi(t) = \frac{\sin 2\pi(t - 1/2)}{2\pi(t - 1/2)} - \frac{\sin \pi(t - 1/2)}{\pi(t - 1/2)}, \tag{E.13}$$

- Ricker (Mexican Hat) wavelets:

$$\psi(t) = (1 - t^2)e^{-t^2/2}. \tag{E.14}$$

E.3 Filter Banks

Practical implementations of the discrete wavelet transform typically use a filter bank in conjunction with resampling blocks to analyze data or signals [51, 52]. Figures E.3 and E.4 show the basic arrangements to perform one-level wavelet decomposition and reconstruction, respectively.

In Fig. E.3, $\bar{h}[n]$ is the impulse response of a lowpass filter and $\bar{g}[n]$ the impulse response of a highpass filter. Those filters split the frequency information of a_j in two. The output of each filter is then downsampled by a factor of 2, i.e., one sample out of two is discarded. At the right side this basic block, we have two signals, a_{j+1}

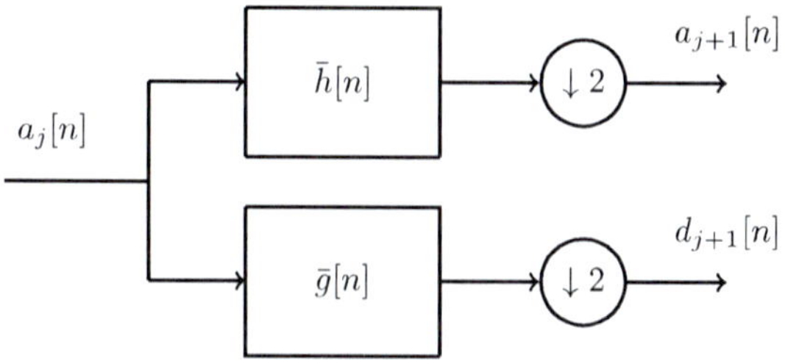

Fig. E.3 Basic block for one-level wavelet decomposition by using a filter bank. The symbol $\downarrow 2$ indicates the downsampling operation by a factor of two

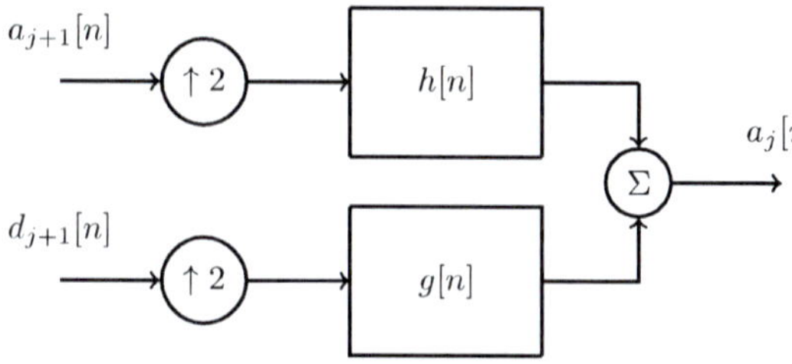

Fig. E.4 Basic block for one-level wavelet reconstruction by using a filter bank. The symbol $\uparrow 2$ indicates the upsampling operation by a factor of two

and d_{j+1}, each with half of the size of the input signal a_j; a_{j+1} and d_{j+1} are referred to as *approximation* and *detail* components of a_j. The next level of decomposition is achieved by feeding the basic decomposition block with a_{j+1}, thus yielding a_{j+2} and d_{j+2}. The process continues until the desired level of decomposition is attained.

At this point, we recall Eq. (E.8) in Sect. E.1, where the approximation of f in $\mathbf{V}_j$ can be expressed as the sum of two functions in subspaces at the next scale $\mathbf{V}_{j+1}$ and $\mathbf{W}_{j+1}$. Specifically, a_{j+1} and d_{j+1} correspond to the inner-products in Eqs. (E.9) and (E.10), respectively.

Wavelet reconstruction is the inverse process, where a_{j+1} and d_{j+1} are first interpolated—upsampled by a factor of 2 and filtered—and then added to recover a_j (Fig. E.4).

In the examples of Chap. 6, we use the wavelet Daubechies-16. The Daubechies wavelets are characterized by having minimum support size for a given number of vanishing moments [53]. In general, there is a trade-off between support size and the number of vanishing moments for a given wavelet [51]. On one hand, the support size influences the order of the filters used to implement a wavelet decomposition. The higher the order, the higher the cost of implementation, in terms of memory, computational effort, and time. On the other hand, the number of vanishing moments is related to the capacity of a wavelet to provide a sparse representation, i.e., to decompose a function into a small number of coefficients. Table E.1 displays the coefficients of the wavelet Daubechies-16 for the filters of the decomposition block. The reconstruction filters in Fig. E.4 are obtained by reversing the decomposition filters:

$$h[n] = \bar{h}[31 - n], \tag{E.15}$$

$$g[n] = \bar{g}[31 - n]. \tag{E.16}$$

Table E.1 Coefficients of wavelet Daubechies 16

n	$\bar{h}[n]$	$\bar{g}[n]$
0	−2.1093396300980412e-08	−0.0031892209253436892
1	2.3087840868545578e-07	0.03490771432362905
2	−7.363656785441815e-07	−0.1650642834886438
3	−1.0435713423102517e-06	0.43031272284545874
4	1.133660866126152e-05	−0.6373563320829833
5	−1.394566898819319e-05	0.44029025688580486
6	−6.103596621404321e-05	0.08975108940236352
7	0.00017478724522506327	−0.3270633105274758
8	0.00011424152003843815	0.02791820813292813
9	−0.0009410217493585433	0.21119069394696974
10	0.00040789698084934395	−0.027340263752899923
11	0.00312802338120381	−0.13238830556335474
12	−0.0036442796214883506	0.006239722752156254
13	−0.006990014563390751	0.07592423604445779
14	0.013993768859843242	0.007588974368642594
15	0.010297659641009963	−0.036888397691556774
16	−0.036888397691556774	−0.010297659641009963
17	−0.007588974368642594	0.013993768859843242
18	0.07592423604445779	0.006990014563390751
19	−0.006239722752156254	−0.0036442796214883506
20	−0.13238830556335474	−0.00312802338120381
21	0.027340263752899923	0.00040789698084934395
22	0.21119069394696974	0.0009410217493585433
23	−0.02791820813292813	0.00011424152003843815
24	−0.3270633105274758	−0.00017478724522506327
25	−0.08975108940236352	−6.103596621404321e-05
26	0.44029025688580486	1.394566898819319e-05
27	0.6373563320829833	1.133660866126152e-05
28	0.43031272284545874	1.0435713423102517e-06
29	0.1650642834886438	−7.363656785441815e-07
30	0.03490771432362905	−2.3087840868545578e-07
31	0.0031892209253436892	−2.1093396300980412e-08

E.4 Thresholding of Wavelet Coefficients

A powerful denoising technique involves the application of thresholding to wavelet coefficients. This process entails the elimination of wavelet coefficients below a specified threshold value, followed by various treatments applied to the remaining coefficients based on the chosen approach to the procedure. The rationale behind this is that the contribution of noise in the wavelet domain is considered to be relatively small compared to the signal terms. However, it should be noted that the optimal threshold for denoising may vary from one sub-band to another. Let's now turn our attention to the two primary thresholding techniques: hard and soft thresholding.

In *hard thresholding*, a wavelet coefficient is preserved as it is, in the case that its absolute value is greater than the threshold; otherwise, the coefficient is zeroed. The procedure is characterized by Eq. (E.17):

$$\hat{d} = \begin{cases} d, & \text{for } |d| > \lambda, \\ 0, & \text{for } |d| \leq \lambda, \end{cases} \tag{E.17}$$

where λ is the threshold, d is the input wavelet coefficient, and $\hat{d}$ denotes the output coefficient.

In *soft thresholding*, however, output wavelet coefficients are determined differently: instead of being preserved when absolute values are above threshold, wavelet coefficients are shrunk, i.e., re-scaled in a way to smooth transition between coefficient values before and after thresholding [54]. Equation (E.18) describes the procedure:

$$\hat{d} = \begin{cases} d - \lambda & \text{for } d > \lambda, \\ d + \lambda & \text{for } d < -\lambda, \\ 0, & \text{for } |d| \leq \lambda. \end{cases} \tag{E.18}$$

Figure E.5a, b illustrate hard and soft threshold functions, respectively.

The most important parameter in this approach is obviously the threshold and the determination of its value can be a true challenge. Better results are achieved by defining optimized values for each decomposition level, as the signal-to-noise ratio varies across sub-bands and coefficient distortion depends on noise characteristics. Donoho [54] calculated optimal thresholds to reconstruct an unknown function from noisy data, where noise is given by independent and identically distributed Gaussian random samples:

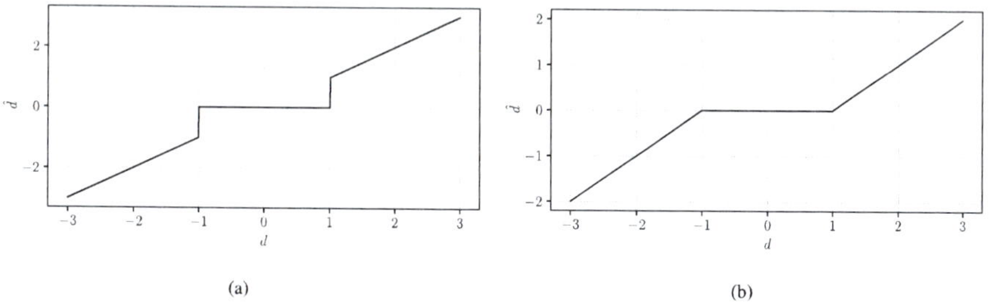

Fig. E.5 Thresholding functions: (**a**) Hard threshold; (**b**) Soft threshold. For both cases, $\lambda = 1$

$$\lambda_j = \frac{m_j}{0.6745}\sqrt{2\ln(L)}, \tag{E.19}$$

where λ_j is the optimum threshold value for decomposition level j; m_j denotes the median values of the wavelet coefficients at level j; and L is the length of the original signal.

Appendix F
Artificial Neural Networks

F.1 A Short Review of Artificial Neural Networks

An artificial neural network (ANN) is a non-linear signal processing system formed by interconnected elements called *artificial neurons*. These neurons are inspired by the structure and operation of biological neurons, allowing ANNs to partially mimic the function of the human brain [55]. Due to their capacity to learn complex patterns, ANNs are widely employed in general classification, regression, and signal representation tasks.

The historical roots of the field of neural networks can be traced back to the seminal works on the theories of biological learning by Hebb [56], and the development of the first mathematical model of a neuron by McCulloch and Pitts [57] in the 1940s. In their model, the authors use propositional logic to describe the activity of a neuron. The model assumes that the input signals to a neuron are weighted to indicate their relative importance, and if the weighted sum of the inputs surpasses a threshold value, an output signal is produced; otherwise, the neuron remains inactive. Another breakthrough in the area was the development of the Perceptron by Rosenblatt in 1958 [58], which is considered the first successful neural network capable of learning and pattern recognition.

In 1969, studies by Minsky and Papert [59] showed the limitations of linear models, which resulted in a significant decline of interest in neural networks. It was not until the development of the backpropagation algorithm by Rumelhart in 1986 [60] that interest in neural networks was resumed. The backpropagation algorithm allowed for the training of multilayer perceptron architectures, which had previously been difficult due to the computational complexity of adjusting the weights in a network with multiple layers. With the backpropagation algorithm, practical applications in image and speech recognition became viable, and neural networks regained their status as an active area of research.

V. Dmitriev et al., *Partial Discharges in Hydroelectric Generators*, Power Systems,
https://doi.org/10.1007/978-3-031-36604-8

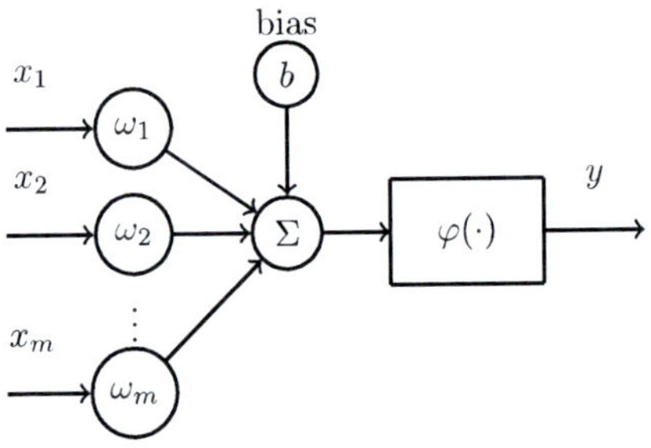

Fig. F.1 Mathematical model of an artificial neuron

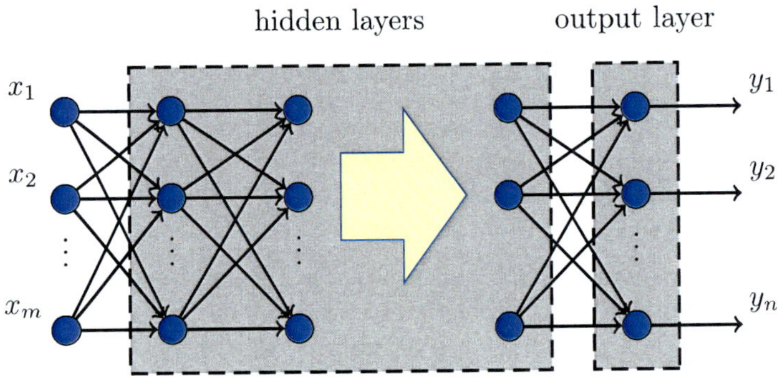

Fig. F.2 Generic multilayer feedforward neural network; each blue circle corresponds to an artificial neuron

The basic processing unit of an ANN is the artificial neuron, shown in Fig. F.1. The neuron applies the following mathematical operations in the inputs x_i to calculate the output y:

$$y = \varphi \left[\sum_i (\omega_i x_i) + b \right], \tag{F.1}$$

where ω_i are the synaptic weights; b is the bias, which serves to increase or decrease the activation function input; and φ is the activation function, which limits the neuron output to a predetermined range [55].

Multiple neurons are linked together in different architectures to enable the learning of complex relationships between inputs and outputs. The architecture of interest in this work is the *multilayer feedforward* [55, 61, 62], shown in Fig. F.2. A neural network learns the relation between inputs and outputs by means of an iterative tuning of weights during the training phase.

F.2 Training, Validation, and Test of Neural Networks

In supervised learning,[3] the synaptic weights of a neural network are trained using a labeled dataset, where the corresponding outputs from input data are known in advance. Such a labeled dataset is randomly split into two major subsets: the training and the test set. The training data are then presented to the ANN, and its weights and biases are updated iteratively. Several algorithms can be employed to this purpose: gradient descent, backpropagation, Adam, and more [62]. In Chap. 6, we use the Scaled Conjugate Gradient Backpropagation algorithm (SCG) [63] (a member of the gradient descent family). The error between the ANN output and labels of the training set, known as the *training error*, decreases with each training iteration.

After training, it is important to evaluate the generalization of the ANN—its ability to perform well on unseen inputs—and this is done with the test data. At some point of the training, the ANN starts becoming excessively specialized on the training data. This common problem in machine learning—known as *overfitting*—is characterized by great disparity between training and generalization errors. Regularization strategies, early stopping procedures, and cross-validation are techniques commonly used to avoid overfitting [62].

F.3 K-fold Cross-Validation

Assessment of performance is of fundamental importance in machine learning science [64]. K-fold cross-validation (CV) allows for comparing different models and selecting the most suitable one for a given problem. In general terms, the best model is the one that better generalizes the problem. Generalization is the ability of correctly and reliably process samples that were not used in training. When a dataset is limited in number of examples, K-fold cross-validation is the most used statistical technique to estimate a model performance [64].

In CV, illustrated in Fig. F.3, the dataset is randomly divided into k mutually exclusive subsets (folds) using stratified sampling, that is, in a way the proportion of classes[4] in each fold is approximately the same of the whole dataset. One fold is taken as the test set, while the others are used for training (or training and validation, depending on the training algorithm). The error over the test fold is calculated. This process is repeated k times, so that each fold has been used for testing once. The overall error estimate is the average of the k test errors.

[3]There are other approaches: semi-supervised and non-supervised learning. As in Chap. 6, our examples use only supervised learning, we decided to keep the focus on this strategy.

[4]The discussion in this section considers the ANN is used as a classifier.

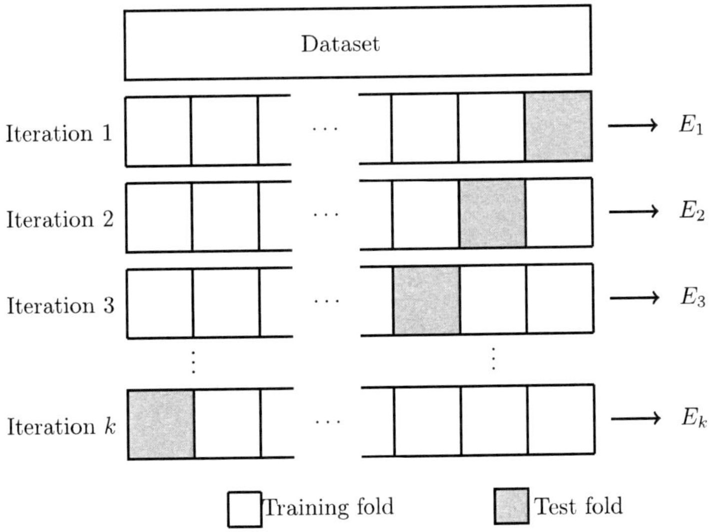

Fig. F.3 K-fold cross-validation. E_i is the error of the i-th iteration

References

1. Stone, G. C. (2012). A perspective on online partial discharge monitoring for assessment of the condition of rotating machine stator winding insulation. *IEEE Electrical Insulation Magazine, 28*(5), 8–13.
2. International Electrotechnical Commission. (2000). High-voltage test techniques-partial discharge measurements. IEC, Publication-60270.
3. International Electrotechnical Commission. (2012). Rotating electrical machines - Part 27: Off-line partial discharge measurements on the stator winding insulation of rotating electrical machines. IEC/TS 60034-27. Technical report. International Electrotechnical Commission.
4. International Electrotechnical Commission. (2012). Rotating Electrical Machines - Part 27-2: On-Line Partial Discharge Measurements on the Stator Winding Insulation of Rotating Electrical Machines. IEC/TS 60034-27-2. Technical report. International Electrotechnical Commission.
5. Institute of Electrical and Electronics Engineers. (March 2015). IEEE Draft Standard for Metal-Clad Switchgear (pp. 1–84). IEEE PC37.20.2/D14.
6. Institute of Electrical and Electronics Engineers. (1997). IEEE Recommended Practice for Voltage-Endurance Testing of Form-Wound Bars and Coils (pp. 1–20). IEEE Std 1043-1996.
7. Institute of Electrical and Electronics Engineers. (1987). IEEE Guide for the Statistical Analysis of Electrical Insulation Voltage Endurance Data (pp. 1–28). ANSI/IEEE Std 930-1987.
8. Institute of Electrical and Electronics Engineers. (1987). IEEE/ANSI Standard for Metal-Clad and Station-Type Cubicle Switchgear (pp. 1–47). ANSI/IEEE Std C37.20.2-1987.
9. Institute of Electrical and Electronics Engineers. (2012). IEEE Recommended Practice for Quality Control Testing of External Discharges on Stator Coils, Bars, and Windings (pp. 1–48). IEEE Std 1799-2012.
10. Institute of Electrical and Electronics Engineers. (2014). IEEE Guide for the Measurement of Partial Discharges in AC Electric Machinery (pp. 1–89). IEEE Std 1434-2014 (Revision of IEEE Std 1434-2000).

11. McDermid, W. (2012). IEEE guide for the measurement of partial discharges in AC electric machinery. In *2012 IEEE International Symposium on Electrical Insulation* (pp. 659–660).
12. Institute of Electrical and Electronics Engineers. (1993). IEEE Standard Definitions of Terms for Antennas. IEEE Std 145-1993 (pp. 1–32).
13. Agência Nacional de Telecomunicações Anatel. (2016). Resolução n° 610, de 18 de abril de 2013. 3.
14. Balanis, C. A. (2016). *Antenna theory: Analysis and design* (4th ed.). Wiley.
15. Sánchez, R. A. (2014). Medida de descargas parciales en radiofrecuencia, tese de doutorado, universidad carlos iii de madri, departamento de ingeneríea eléctrica.
16. Harlow, F. H., & Welsh, J. E. (1965). Numerical calculation of time-dependent viscous incompressible flow of fluid with free surface. *Physics of Fluids, 8*, 2182–2189.
17. Yee, K. (1966). Numerical solution of initial boundary value problems involving Maxwell's equations in isotropic media. *IEEE Transactions on Antennas and Propagation, 14*, 302–307.
18. Taflove, A., & Brodwin, M. E. (1975). Numerical solution of steady-state electromagnetic scattering problems using the time-dependent Maxwell's equations. *IEEE Transactions on Microwave Theory and Techniques, 23*, 623–630.
19. Taflove, A., & Brodwin, M. E. (1975). Computation of the electromagnetic fields and induced temperatures within a model of the microwave-irradiated human eye. *IEEE Transactions on Microwave Theory and Techniques, 23*, 888–896.
20. Taflove, A. (1980). Application of the finite-difference time-domain method to sinusoidal steady-state electromagnetic penetration problems. *IEEE Transactions on Electromagnetic Compatibility, 22*, 191–202.
21. Teixeira, F. L. (2022). A summary review on 25 years of progress and future challenges in FDTD and FETD techniques. *The Applied Computational Electromagnetics Society Journal, 25*(1), 1–14.
22. Bayliss, A., & Turkel, E. (1980). Radiation boundary conditions for wave-like equations. *Communications on Pure and Applied Mathematics, 23*, 707–725.
23. Mür, G. (1981). Absorbing boundary conditions for the finite-difference approximation of the time-domain electromagnetic field equations. *IEEE Transactions on Electromagnetic Compatibility, 23*, 377–382.
24. Higdon, R. (1987). Numerical absorbing boundary conditions for the wave equation. *Mathematics of Computation, 49*, 65–90.
25. Liao, Z., Wong, H., Yang, B. P., & Yuan, Y. F. (1984). A transmitting boundary for transient wave analysis. *Scientia Sinica, XXVII*(series A), 1063–1076.
26. Berenger, J. (1994). A perfectly matched layer for the absorption of electromagnetic waves. *Journal of Computational Physics, 114*, 185–200.
27. Sacks, Z., Kingsland, D., Lee, R., & Lee, J. (1995). A perfectly matched anisotropic absorber for use as an absorbing boundary condition. *IEEE Transactions on Antennas and Propagation, 43*, 1460–1463.
28. Holland, R. (1983). Finite-difference solution of Maxwell's equations in generalized nonorthogonal coordinates. *IEEE Transactions on Nuclear Science, 30*, 4589–4591.
29. Mei, K. K., Cangellaris, A., & Angelakos, D. J. (1984). Conformal time domain finite-difference time-domain method. *Radio Science, 19*, 1145–1147.
30. Lee, J. F., Palendech, R., & Mittra, R. (1992). Modeling three-dimensional discontinuities in waveguides using nonorthogonal FDTD algorithm. *IEEE Transactions on Microwave Theory and Techniques, 40*, 346–352.
31. Noda, T., & Yokoyama, S. (2002). Thin wire representation in finite difference time domain surge simulation. *IEEE Transactions on Power Delivery, 17*, 840–847.
32. Baba, N. N. Y., & Ametani, A. (2005). Modeling of thin wires in a lossy medium for FDTD simulations. *IEEE Transactions on Electromagnetic Compatibility, 47*, 54–60.
33. Zhang, X., Fang, J., Mei, K. K., & Liu, Y. (1988). Calculation of dispersive characteristics of microstrips by the time-domain finite-difference method. *IEEE Transactions on Microwave Theory and Techniques, 36*, 263–267.

34. Fornberg, P. E., Byers, A., Piket-May, M., & Holloway, C. L. (2000). FDTD modeling of printed circuit board signal integrity and radiation. In *IEEE International Symposium on Electromagnetic Compatibility*.
35. Müller, F. C. B. F., Farias, R. G., Sobrinho, C., & Dmitriev, V. (2005). Multistatic radar with ultra wideband pulses: FDTD simulation. In *International Microwave and Optoelectronic Conference*, Brazil.
36. Goorjian, P. M., & Taflove, A. (1992). Direct time integration of Maxwell's equations in nonlinear dispersive media for propagation and scattering of femtosecond electromagnetic solitons. *Optics Letters, 17*, 180–182.
37. Kondylis, G., DeFlaviis, F., & Pottie, G. (1999). Indoor channel characterization for wireless communications using reduced finite difference time domain (R-FDTD). IEEE VTC Fall 99, Amsterdam.
38. Maloney, J. G., & Kesler, M. P. (1998). Analysis of periodic structures. In A. Taflove (Ed.), *Advances in computational Electrodynamics*. Artech House.
39. Oliveira, R. M. S., & Sobrinho, C. (2009). Computational environment for simulating lightning strokes in a power substation by finite-difference time-domain method. *IEEE Transactions on Electromagnetic Compatibility, 51*(4), 995–1000.
40. de Oliveira, R. M. S., Modesto, J. F. M., Dmitriev, V., Brasil, F., & Vilhena, P. (2016). Spectral method for localization of multiple partial discharges in dielectric insulation of hydro-generator coils: Simulation and experimental results. *Journal of Microwaves, Optoelectronics and Electromagnetic Applications, 15*, 170–190.
41. de Oliveira, R. M. S., Rodrigues, N. R. N. M., & Dmitriev, V. (2015). FDTD formulation for graphene modeling based on piecewise linear recursive convolution and thin material sheets techniques. *IEEE Antennas and Wireless Propagation Letters, 14*, 767–770.
42. Manteuffell, D., & Simon, W. (2005). FDTD calculation of the SAR induced in the human head by mobile phones: New standards and procedures for the numerical assessment. In *IWAT 2005 (IEEE International Workshop on Antenna Technology)* (pp. 137–140).
43. Taflove, A., & Hagness, S. C. (2005). *Computational electrodynamics, The finite-difference time-domain method* (3rd ed.). Artech House.
44. Jackson, J. D. (1962). *Classical electrodynamics*. New York: Wiley.
45. Gedney, S. D. (1996). An anisotropic perfectly matched layer absorbing media for the truncation of FDTD latices. *IEEE Transactions on Antennas and Propagation, 44*, 1630–1639.
46. Daryanani, G. (2009). *Principles of active network synthesis and design*. Wiley India Pvt. Limited.
47. Oppenheim, A., & Schafer, R. (2013). *Discrete-time signal processing: Pearson new international edition*. Pearson Education.
48. Parks, T., & McClellan, J. (1972). Chebyshev approximation for nonrecursive digital filters with linear phase. *IEEE Transactions on Circuit Theory, 19*(2), 189–194.
49. Mix, D. F., & Olejniczak, K. J. (2003). *Elements of wavelets for engineers and scientists*. Wiley.
50. Akujuobi, C. M. (2022). *Wavelets and wavelet transform systems and their applications*. Springer International Publishing.
51. Mallat, S. (2009). *A wavelet tour of signal processing: The sparse way*. Academic Press.
52. Vetterli, M., & Kovačevic, J. (1995). *Wavelets and subband coding*. Prentice-Hall, Inc.
53. Daubechies, I. (1992). *Ten lectures on wavelets*. Society for Industrial and Applied Mathematics.
54. Donoho, D. (1995). De-noising by soft-thresholding. *IEEE Transactions on Information Theory, 41*(3), 613–627.
55. Haykin, S. (2004). *Neural networks - A comprehensive foundation*. Prentice Hall.
56. Hebb, D. O. (1949). *The organization of behavior: A neuropsychological theory*. New York: Wiley.
57. McCulloch, W. S., & Pitts, W. (1943). A logical calculus of the ideas immanent in nervous activity. *Bulletin of Mathematical Biophysics, 5*, 115–133.

58. Rosenblatt, F. (1958). The perceptron: A probabilistic model for information storage and organization in the brain. *Psychological Review, 65*(6), 386–408.
59. Minsky, M., & Papert, S. A. (1969). *Perceptrons*. The MIT Press.
60. Rumelhart, D. E., Hinton, G. E., & Williams, R. J. (1986). Learning representations by back-propagating errors. *Nature, 323*, 533–536.
61. Aggarwal, C. C. (2018). *Neural networks and deep learning*. Springer International Publishing.
62. Goodfellow, I., Bengio, Y., & Courville, A. (2016). *Deep learning*. MIT Press.
63. Møller, M. F. (1993). A scaled conjugate gradient algorithm for fast supervised learning. *Neural Networks, 6*(4), 525–533.
64. Witten, I., Frank, E., & Hall, M. (2011). *Data mining: Practical machine learning tools and techniques*. Morgan Kaufmann.

Index

V. Dmitriev et al., *Partial Discharges in Hydroelectric Generators*, Power Systems,
https://doi.org/10.1007/978-3-031-36604-8